HIRAM PERCY MAXIM

Father of Amateur Radio,
Car Builder
and
Inventor

by

Alice Clink Schumacher

The Ham Radio Publishing Group
Greenville, N.H. 03048

Printed in the
United States of America
All Rights Reserved
ISBN: 0 918232-04-X
Library of Congress Card Catalog Number 76-057438

ACKNOWLEDGMENT

Material from Hiram Percy Maxim's **Horseless Carriage Days** (c. 1936 and 1937), and **A Genius in the Family** (c. 1936), is used through the permission of the publishers, Harper and Rowe.

Appreciation is also expressed to the Dover Publications, who did a reprint of Maxim's books (c. 1962) under special arrangements with Harpers.

Acknowledgment is further made to the American Radio Relay League, Inc. of Newington, Connecticut, for granting permission to use passages from "Two Hundred Meters and Down," by Clinton B. DeSoto, and material from early-day QST's, including the poem "On February 17th," by Michael J. Caveney, VE3GG.

DEDICATION

To my son, Peter Schumacher - WØJYJ
(Ex KR6SO -- KG6FAE -- and K7MOY)

With special thanks to:

PERCY MAXIM LEE

John Glessner Lee
Hiram Hamilton Maxim
John Maxim Lee
Roland Bourne
Charles Barrett
J. A. Sundkvist

and

The Staff of the American Radio Relay League
The Connecticut State Library at Hartford
The Great Falls (Montana) Public Library

FOREWORD

When Alice Schumacher first informed me of her intent to write a book about Hiram Percy Maxim, my father, I felt sure no amount of reading or talking with people who had known him would be as meaningful to her as would a study of the home movies in which he appeared. The quality of his mind was revealed in what he wrote and what he achieved. His wit came through his lectures and stories. The sincere devotion of his associates was a story in itself. A clear record of his character was expressed in the conduct of his life. In the movies of him these qualities are apparent, but his unique and appealing personality dominates. His gestures, his smile, the way he threw back his head when he laughed, his intense concentration on subjects which interested him, his kindliness and sincerity, his love of life - these are the things one senses when he is seen in motion. This is how Mrs. Schumacher has caught his spirit.

Undoubtedly the activity which gave HPM the greatest satisfaction was his work in Amateur Radio. The ingredients that challenged him were the association with young men of modest or no means; their ingenuity and determination; their sense of high adventure. He once wrote: "the hardest working among us are those who work with their brains. The driving force seems to be our own incurable ambition to be somebody —to accomplish something." Amateur Radio, though a hobby for him, gave him a chance to help others in this forward movement and provided him with associations which he valued above all others. And as he said on so many occasions: "73 to my fellow amateurs!"

(Signed)
Percy Maxim Lee
Farmington, Connecticut

1/19/70

HPM wearing arm band in mourning
for his mother, who died in 1911

INTRODUCTION

Who was Hiram Percy Maxim?

The motor car enthusiast could say:
1. He was the man responsible for the steering wheels of American automobiles being on the left side instead of on the right.
2. He won the first track car race ever held in the United States.

The historian could say:
1. His father was a famous inventor, knighted by Queen Victoria.
2. He was a graduate of Massachusetts Institute of Technology at sixteen, and a practicing engineer at seventeen.

The scientist could say:
1. He pioneered in gliders, and in the field of air cleaning and air cooling.
2. He pioneered in radio, and foresaw TV.
3. He participated in America's first overseas radio communication (1921).
4. He was an expert in planetary observation, and foretold the potential of the space age.
5. He invented the famous Maxim gun silencer.

The society commentator could say:
1. He was married to the daughter of a former Governor of Maryland.
2. He was known internationally... to statesmen, inventors, authors, governors, presidents and scientists as "The beloved gentleman."

The humanitarian could say:
1. He realized that noise, a by-product of modern civilization, was harmful, and took steps to curtail it.
2. He was the **father** and savior of **amateur radio**, which has helped so many thousands in times of emergency and disaster, and which has been a constructive hobby for men, women and children for decades.

Men who DO things have a motivating word. Hiram Percy Maxim had two .... **organization** and **unselfishness.**

Taken april 23, 1908. H.P.Maxim
Hartford, Conn.

Left to right: The first Maxim Silencer, on a Winchester rifle. Hiram Percy Maxim, Hiram Hamilton Maxim (8 years), Mr. Jenkins (the patent attorney), T. Wells Goodridge (partner and HPM's best man), and Simon Britt (the machinist who made the silencer). Taken April 23, 1908.

HPM with one of his horseless carriages - 1898

Hiram Percy Maxim

When the Maxim family fled France early in the 17th century, they were to leave their mark on our modern life all the way from cars, munitions and noise control, to air conditioning and communications.

Family tradition tells it this way: "When Isaac Maxim, son of sea captain Samuel Maxim, married the daughter of old Brimstone Stevens, they give rise to seethin' torrents."

The Maxims were French Hugenots who emigrated from France about 1645 in order to escape religious persecution. They landed near what is now Wareham, but after a time moved on to Rochester, Massachusetts, where they settled. Eventually their children located in Maine, where their descendants lived for the next three centuries.

Isaac Maxim, born in 1818, married tiny but dynamic Harriet Boston Stevens, when she was 23. During their marriage they moved 33 times.

In the inland village of Wayne, Maine, the old Maxim house still stands, between the center of town and the cemetery. Once white, with a metal roof, the one-story gabled structure boasts a fine New England doorway with glass side panels and a fanlight. The wood carving above it features a design of stars, executed by some early Maxim. The stars are prophetic.

Isaac and Harriet Maxim reared several sons. One of these was the rather violent Henry, renowned in his childhood for biting a goose to death after it had attacked him first, and for killing a vicious forty-pound dog single-handed in the crawl space under a general store in Abbott Village. Henry became a soldier in the Civil War, but he got sick and was sent home, where he died in bed.

Another brother was Isaac Maxim, who adopted the name Hudson and was to become famous as the inventor of smokeless powder and other creations, some of which were sold to the DuPonts. Less publicized, was his appearance in a parade as King Neptune at an Atlantic City Beauty Pageant.

The most famous of the brothers was Hiram Stevens Maxim, born in 1840 in Sangerville, Maine. He epitomized Yankee ingenuity and eccentricity. Although he had less than five years of formal schooling, he studied science and engineering while traveling around the country working at odd jobs. He eventually became a famous inventor, scientist and engineer, and the head of his own company. He married Jane Budden, and their first child and only son was born in Brooklyn on September 2, 1869.

Hiram Percy Maxim came into a world that was still without electric lights, power streetcars, bicycles, automobiles, telephones or recorded sound. Travel was on foot, on horseback, or by rail. Airplanes and radio were only fantastic dreams. However, the Englishman, Faraday, had established the principle of radiation and magnetic induction, and had founded the electromagnetic theory of light. And on a mountain in West Virginia a group of men had sent up a kite with a square of copper gauze and a trailing wire. At the base of another mountain 18 miles away was a second and similar kite. A Washington D.C. dentist, Dr. Mahlon Loomis, activated a galvanometer connected with a buried coil of wire. The other galvanometer 18 miles away quivered. This was the stir of life at the birth of wireless communication, and coincided with the birth of that Maxim who was to become radio's mentor.

When Hiram Percy Maxim was four years old he lived in Brooklyn, on 3rd Street, near Smith, and close to the horsecar line. The house had a brownstone stoop by which one entered the second floor, with its parlor, living room and bedrooms. The kitchen, dining room and pantry were below, with a basement entrance. A fancy iron fence surrounded a sickly lawn, made that way by too much boyish activity. Little Percy was always in a hurry. He ran instead of walked. He took stairs two and three at a time. He tore through doors and gates and left them swinging open. His impetuous speed kept him from stopping properly, and he was always crashing into something or somebody. He was full of life, and intensely curious about it. Moreover, he had an unusual childhood because he had an unusual father.

Hiram Stevens Maxim was then an inventor, the senior partner in the firm of Maxim and Welch, makers of steam engines and gas generating machines. Their plant was on Center Street in New York City, and Maxim Senior commuted daily. Maxim Junior had a very patient mother, and a baby sister, Florence, who cried a lot. Under such circumstances, it would be nice if a boy had a dog.

This desire for a pet led to the incident of the two-headed penny. In 1873 the prime purpose of a drug store was to dispense drugs, and the one in the Maxim neighborhood was old-fashioned, even for those days. It was dazzling with fancy glass bottles filled with colored liquids. But more important to young Percy Maxim, the proprietor owned a little white dog. Every time the boy went into the establishment on an errand he came to love the animal more, and to desire it above anything in the world. Dallying at the counter one day after his purchase was completed, the little four-year old told the druggist how much he loved the dog, and that he would like to have it. This scene was repeated with frequency and intensity for a number of days. Finally the druggist suggested a deal.

"Bring me a penny with a head on each side," he told the child, "and I'll give you the little white dog."

It was a bargain he would never have suggested had he realized that he was making the offer to the son of one of the most brilliant practical jokers of the century.

The boy raced home and told his mother of the wonderful opportunity. Mrs. Maxim, already hardened to such trickery by her husband's elaborate pranks, told her young son that he was wasting his breath, that there simply was no such thing as a two-headed penny. After a fruitless search through his mother's purse, young Percy came to the conclusion that such matters were above a woman's head anyway, and he decided to appeal to his father, because he could do great things.

Strictly speaking, Hiram Stevens Maxim never acted much like a father. He did strange and exciting things like vaulting the cast iron fence when he came home from work, seemingly unhampered by his high silk hat and Prince Albert coat, or the stares of the neighbors.

So, on that long-ago summer night, Percy waited for his father to return from his factory. When he saw him coming, he ran to meet him, breathlessly explaining about the little white dog that the druggist had promised to give him, as soon as he brought in a two-headed penny.

Maxim Senior was weary with the problems of a long and important day, but he stopped and listened. Soberly he searched through his pockets for pennies, and carefully examined those he found. None of them had two heads. However, he appeared undaunted. "Never mind," he told his disappointed son. "I'll bring you one from New York tomorrow. They must have plenty of them over there."

Young Percy was content. He knew that nothing was impossible for his father. All he had to do was wait until the next night and the little white dog would be his.

The next day dragged frustratingly. All the details were planned—where the dog would sleep, and what it would eat.

Meanwhile, the senior partner in the firm of Maxim and Welch had had a busy day in the factory tool room. He had put a penny in a lathe and faced off the tail side until it was exactly half the thickness of a real penny. Then he repeated the process with another, soldered the two halves together, and painstakingly burnished the joint. He dropped the finished coin in his pocket along with his other small change.

That night when his son ran to meet him, Maxim Senior conducted an elaborate search, finally coming up nonchantly with the coveted penny. Percy was all for rushing off to the drug store immediately, but his father insisted that he wait until after supper, when he would accompany him. It was a confused meal, with Mrs. Maxim breaking into the conversation when she could, begging her husband "not to do anything foolish, or in bad taste." The boy wondered about all the talk. What possible trouble could there be?

When the current one of their succession of hired girls came to clear off the table, Maxim Senior gave the approving signal, and father and son set off for the drug store. Percy rushed up to the astonished druggist and thrust the two-headed penny at him.

"Here it is!" he exclaimed. "I brought it. Now I can have the dog!"

The druggist turned the coin over and over in his fingers, as Maxim Senior watched quietly.

"Where did you get this?" he demanded.

"From my father," Percy shrilled. "Now I want my dog. You promised!"

As soon as the bewildered shopkeeper recovered his faculties, he began to argue. No such coin had ever been minted, he was certain. And he had no intention of giving up the little white dog.

When he had finished, and while Percy manfully fought back tears, Maxim Senior proceeded to give the druggist the lecture of his career. He finished by making it plain that in the future it would be more prudent not to make any offers unless he intended to live up to them.

It was Percy's first experience with a broken promise, and one of his first lessons in growing up. After that, there were many more interesting adventures with his father. On one occasion, Maxim Senior learned that cattle in the stockyards were being kept without any water, so that on sale day they would drink huge quantities and thus weigh more, and bring in more money to their owners. He was so horrified at this cruelty that he walked out into the yards and opened all the water valves for the thirsty beasts. Then he reported the inhumane treatment to the proper officials and was instrumental in having a law passed to prevent any further such practices.

Not all of his dramatic actions were so well motivated.

He once delighted the children, and horrified his wife, by performing an experiment in the kitchen. This involved one of the maids, and its purpose was to show that the sensations of extreme heat and extreme cold are indistinguishable. He had made a big show of heating up a poker until it was white hot, and then placing it in a bucket beside the kitchen stove. The next time the unsuspecting maid walked past, he reached over from behind and touched her neck with a piece of ice. She promptly flew into hysterics, screaming that she had been branded. Even when they showed her in a mirror that there was not the slightest mark, she continued to wail, and the Maxims lost another maid.

Although it was during this period that Mr. and Mrs. Maxim went occasionally to hear Henry Ward Beecher preach, Percy received little formal religious training. The family did, however, have a high church neighbor appropriately named Mrs. Righter, and she had undertaken to teach Percy a few fundamentals, including the Ten Commandments. Maxim Senior was in the habit of taking his son fishing with him on Sunday. This was considered a very wicked thing in those days, and Maxim Senior compounded the offense by telling Percy that Mrs. Righter had taught him wrong, that it should be, "Remember the Sabbath and go fishing." When Percy in all innocence ran to inform Mrs. Righter of her theological error, Mrs. Maxim had another diplomatic tangle to unravel.

It was that same year, 1873, that James Clerk Maxwell published a book about the electrical and magnetic theories which would activate sound waves in the ether. The parallel between young Percy and his radio destiny was progressing. Also at this time, a lawyer by the name of William Thomas Hamilton was elected to the United States Senate. He would later become governor of Maryland, and eventually Hiram Percy Maxim's father-in-law.

In the spring of 1875 the Maxim family moved to Fanwood, New Jersey. There they had a horse and carriage, a barn, pigs, chickens and a cow. There was also a garden, a blackberry patch, and a large pond with a clumsy raft. Percy was six years old then, and his birthday present was being taken to school for the first time.

The boy had many unusual memories of that farm. One of them concerned the time his Uncle Frank was painting the house and confused the color spacing, so that the back of the building did not match the sides and front. The resultant loud, poetic blasphemy made Mrs. Maxim ill, terrorized the servant girl, and fascinated young Percy.

One Sunday, Percy thought it would be pleasant to take a ride across the pond on the raft. His father cooperated with the idea, and Percy's mother, his little sister Florence and the hired man assembled to witness the disembarking. Unfortunately the crude craft was awash and barely floating. When his father took a pole and shoved the thing out into the water, Percy became terrified. He dreaded crying, because he knew that it always made his father very angry, but the thought that he was never going to see his dear mother again overcame him, and he broke into tears. So did Mrs. Maxim, and she insisted that her son be brought back. That put both of them in disfavor, and for days young Percy was miserable with the shame of failure.

Two other rural incidents were pleasurable. The first involved the hypnotizing of a chicken. Maxim Senior, interested in the new practice of mesmerizing, decided on an experiment in his barn, and assigned Percy the task of chasing down the chicken. This was unbelievably good news, as he was usually scolded for chasing the hens. After a great deal of wing flapping and feather flying, the fowl was placed on the barn floor, and a chalk mark was drawn straight out from its beak. Percy was delighted to observe that for several minutes the chicken sat in a complete stupor.

The second incident involved "The Seedy One," an untidy unclean itinerant who appeared on the Maxim farm and offered to pray, in return for money for a meal. Maxim Senior invited him into the parlor and told him he would pay him if his prayers could stop the music box. He put on its ten perforated metal discs and was enjoying the competition of the Seedy One's prayers with the jolly tunes of the music box, when Mrs. Maxim put in an unexpected appearance. The experiment on the efficacy of prayer ended suddenly, and the Seedy One got the money for his meal.

During the summer, little Percy became a real companion to his mother when she took him frequently to an ice cream pavillion in Plainfield, New Jersey. They ate and laughed together as they sat overlooking a little lake. In later life, he often remarked that although he had eaten at Monte Carlo, Nice, Cairo and Naples, none of their fares ever compared with the pavillion ice cream he shared with his mother as a boy in New Jersey.

In October, a new little sister, Addie, was born. Her advent initiated Percy into the "babies in the cabbage patch" theory, and all the confusion in the household frightened him into thinking that the house was on fire.

About this time, at the age of seven, Percy received his first glimmer of what it would be like to be a man. The occasion was the Centennial Exhibition in Philadelphia, commemorating the one hundredth anniversary of the signing of the Declaration of Independence. Perhaps he remembered the incident more clearly because of the uproar that ensued when Maxim Senior announced that he was going to attend and take his little son along. Mrs. Maxim was concise in her protests. She foresaw a succession of unfortunate practical jokes, of homesickness, and the possibility of tears from Percy and consequent ridicule from his father. In fact, to her the proposition suggested nothing but trouble. Finally, however, they won her over, and hand in hand, father and son toured the exhibition grounds and displays. To Percy it was like walking through a new world. And he was fortunate enough to have a father who could explain everything, in language that a boy could understand.

Percy was entranced by a little steam engine, and insisted on coming back to it again and again. But most of all, he was impressed by a huge meteorite. The sign said that it weighed 1440 pounds, and Maxim Senior explained that it had fallen to our earth from the sky—that it had probably sailed in space for millions of centuries, and was not of our earth. The feeling of reverence and awe that overwhelmed Percy at these words was to inspire him with searching thoughts for the rest of his life.

Two years later, in 1877, the family moved back to Brooklyn, into a brownstone house on Union Street. The front stoop led up into a living room, dining room, and butler's pantry. This latter featured a dumbwaiter which connected with the kitchen directly below. There was also a supply closet, where the children were put when they misbehaved.

Maxim Senior had just organized the United States Electric Lighting Company. As its chief engineer he was working days at the plant, and nights at his drafting table at home. He was trying to develop an incandescent electric lamp. Since scientific experiments did not receive much publicity in those days, he was unaware that Edison was working on the same project. The latter beat Maxim to the patent office with a successful product by only a few days.

The back yard of the Brooklyn home boasted a peach tree, but Percy was unhappy over the fact that there were no peaches on it. Soberly his father explained that in order to grow peaches, one should bury a dead cat at the foot of the tree for fertilizing. The boy spent several days in search of such an animal, and finally discovered one, in an advanced state of decay. He dragged it home, and he and his father interred it beneath the fruit tree. Then he and Florence ran happily off to Sunday School. When he got home, his father called him to come and see the beautiful peaches. It was true. The tree was loaded with them—large juicy fruit, carefully impaled on the branches. Gleefully Percy harvested the crop, probably right back into the same basket in which Maxim Senior had just toted them from the fruit stand.

The incident of the bat which took place a few weeks later was not so pleasant, and it is no wonder that Mrs. Maxim was constantly exclaiming over something her husband had done. There had been some family discussion about bats, and Percy was curious to locate one and study it first hand. He was very excited, therefore, when one day his father shouted, "Percy, come quickly! There's a bat in the dumbwaiter shaft!"

Percy ran to fetch the broom, and his father held him

upside down by the heels as he flailed away. But somehow the bat managed to dodge the blows. Maxim was so overcome with laughter that he nearly dropped his son down the shaft. Finally he lifted him out, dirty and perspiring, and explained that the object wasn't really a bat after all, but a black bow that he had snipped from his wife's hat and suspended into the shaft on a wire spring. That experience left young Percy with another lifelong thought—wonderment that any man would dare to mutilate his wife's hat.

Ordinarily, Percy was a good little boy, and almost never had any serious problems with his conscience. On one occasion, however, he had an unfortunate accident that involved his mother's cherished pier glass. Florence had been amusing herself by bouncing a marble off the base of the mirror and catching it. Percy decided that it would be fun to reach in from behind and snatch the marble. But he misjudged his grab, hit the marble into the mirror, and broke it. He went dutifully and told his mother. She cried, and said she would have his father administer a whipping. Seeing his mother cry was already punishment enough to the sensitive boy, and supper was a sad affair. When it was over, father and son went out into the yard for switches, and then upstairs into a bedroom, where Maxim removed his coat and tie, and rolled up his sleeves. Percy was impressed. The worst physical punishment he could remember was a smart little tap with a T square.

After a few preliminary swishes into the air with the chosen twigs, and mentioning other possible instruments ranging from umbrellas to baseball bats, Maxim settled on the use of his cane. This he proceeded to whack savagely into the bed covers, sending horrible sounds down the stairwell for the benefit of Mrs. Maxim. Finally he explained to Percy that with all his testing he was unable to find a suitable device for whipping, so that this time there would be no punishment. But he cautioned his son that in the future he should be more careful around the house, and should try to be a better boy. The reply was a sincere, "Yes, Papa, I will." He later referred to the incident as his first experience in "engineering research."

This episode was followed shortly by another domestic

crisis. Percy's sister Florence came home from school in tears because her teacher had insinuated that she was mentally defective, due to her total lack of comprehension of arithmetic. Maxim Senior was all for going down and taking the school apart, but thought better of it and determined to reverse the situation. Recruiting Percy as an unsuspecting assistant, he began a nightly series of "games." By clever manipulations of rows of beans, putting some in and easing some out, he had the children gleefully competing in fast calculations of addition, subtraction and multiplication. Florence became the head of her class, no doubt to the wonderment of her teacher.

Since Percy had been the butt of his father's practical jokes ever since he could remember, it was inevitable that eventually he should attempt one himself. And what better victim than that same father? The inspiration struck him one April Fool's Day at school, when an older boy brought a sack of trick candy which produced spectacular results. In reality, these were pieces of laundry soap, cunningly coated with chocolate. After a great deal of coaxing, the little gentleman agreed to give Percy a piece of the candy to take home to his father. Following dinner, which seemed to drag painfully, Percy approached Maxim Senior, who had settled down to read.

"Papa, would you like a piece of candy?" he queried.

There was a measured hesitation, as father regarded son in a lightning inventory of possibilities. Then he nodded calmly, took the proffered sweet and stuffed it into his mouth, or at least appeared to do so. Gleefully the boy waited for the explosion. But nothing happened. Maxim Senior continued to read his paper, chewing calmly. Finally the suspense became intolerable.

"Did you like the candy, Papa?"

"Why yes, Percy. Very much. Do you have any more of them?"

Maxim Junior was shaken down to the very toes of his stout little shoes. For months afterwards he pondered the situation. Had his father really eaten the soapy concoction, or had he managed to slip it out into his hand and dispose of it later? Or was it possible that the boy at school had feared the consequences of having a part in such a daring affair, and given him a piece of good candy? Since it was a subject that he did not care to discuss with his father, it went into the limbo of childhood mysteries.

In 1878, when Percy was nine, he and his father were watching a New York parade in honor of General U. S. Grant. Maxim Senior baited an inebriated man in the crowd by pretending that he did not know who General Grant was. He teasingly identified him as a poet, and as the portly stranger grew more red in the face with argument, Maxim went further, and also disclaimed any knowledge of Abraham Lincoln, George Washington, or Julius Caesar. Fortunately at this juncture a policeman appeared and escorted the intemperate and belligerent man into a nearby hotel.

Another New York incident about this time embarrassed both Mr. and Mrs. Maxim. Percy had once been inside a barroom with his father, and had been entranced with the painting of a man in a striking red cloak. The smiling bearded fellow was seated on a barrel, and holding out a mug of beer. When Percy had inquired who the man was, his father offhandedly remarked that he was the one who invented lager beer. So it came about that when Percy and his parents were walking through the galleries of the Metropolitan Art Museum in New York, the boy was excited to see a gorgeous painting of a man in a red cloak. In a voice that could be heard all over the room Percy proudly announced, "There's the man that invented lager beer!"

Mrs. Maxim was embarrassed again in New York when she and a lady acquaintance attended the theatre, accompanied by little Percy. His pet fear was Indians, and the play they were watching was full of them. In vain the boy nudged his mother, and whispered pleadingly with her to escape while they could. Absorbed in the action of the drama, she tried to quiet him. But when Percy saw a horde of half-naked savages plunging across the stage with their tomahawks upraised, and warpath screams coming from their copper throats, he leaped from his seat and out into the aisle. He ran toward the exit yelling, "Come on, Mama!" She came, but she was very annoyed.

It was during this period that Mr. Maxim, deeply occupied with experiments and projects, became alarmingly absent-minded. He was continually losing track of personal possessions, including important drawings, so that he suffered great inconvenience and annoyance. On one of these occasions he appealed to his young son for help. Enjoying the domestic histrionics, he seemed so tragically upset that the boy actually thought he was going to burst into tears. Anxious to help, he suggested that his father mark with his name and address everything that he carried about with him. Maxim Senior seemed delighted with the idea, and came home in a few days with a box of stickers which he proceeded to glue onto everything he owned, from books to umbrellas. They read: "This was lost by a damned fool named Hiram Stevens Maxim, who lives at 325 Union Street, Brooklyn. A suitable reward will be paid for its return."

In spite of his temperamental outbursts, Mr. Maxim was usually a controlled and even-tempered man. There was only one thing which never failed to make him cross, and that was, as he put it, "disorganization in the writing materials department." He was never upset by noise, broken things, or bawling children, but he was like a madman when there was a supply failure in the line of envelopes, paper, blotters or ink. Unfortunately, Percy was generally the cause of the trouble. His father would bring home great quantities of stationery supplies, and his mother would put them carefully away in the desk. From that moment, temptation took over. When Percy needed a sharp tool with which to pick at something, he used a pen. Of course this damaged the point. Those same versatile points, coated with oil, could be floated in a dish of water and pulled about with a magnet. His father had showed him this fascinating game one rainy afternoon. It was a wonderful experience in attraction and repelling, but hard on the supply of writing materials. Fortunately, Mrs. Maxim did not mind the use of the large silver bowl on the dining room table as a pond for these operations. The service survived this usage, and when the boy became a man he gave it to his wife, who used and

treasured it for thirty-five years. The family penholders did
not fare so well. Percy used them as axles for little wheel-
barrows and carts, or employed them to plug holes in things.
And ink was the only paint the boy ever had. Everything he
made which needed to be colored had to be smeared with ink.
Worst of all, the detestable bottles were always getting tipped
over. This necessitated the use of huge quantities of blotters,
so that their supply was continually exhausted. The envelopes
were useful for many things, and the notepaper was handy for
doing homework. However, this last was such a detestable
occupation that it did not make too much of a demand on the
supply. As Percy himself said in later life, his father used to
scratch around in that desk "like a squirrel hunting for nuts."
When he failed to find what he was looking for, he was apt to
appeal to the Almighty to look down in confirmation that there
was no pen, in spite of the fact that he had just purchased one.
He would groan and carry on, and then finally jam his hat down
over his ears and go out and buy more supplies. In reality,
he enjoyed these little scenes. They helped him to relieve the
pressures brought on by the trying details of being an inventive
genius.

Several times when he was little, Percy had been taken
to Wayne, Maine to visit his paternal grandparents. But during the boy's ninth summer, Mr. Maxim decided he could make
the trip by himself. Mrs. Maxim objected, but was overruled.
The trip seemed like a lark to Percy, until it was underway,
and he found he had to take the Fall River boat from New York
and then proceed by train, with several changes via Boston and
Portland. He was a very homesick little boy when his Uncle
Sam met him at Winthrop, but being allowed to help drive the
horses was some comfort.

One little event of that summer became a part of Percy's
lifelong memories. His grandfather had introduced him to an
elderly man, explaining that he used to be Grandpa's school
teacher. Since the fellow seemed so incredibly old to the boy,
he questioned his grandfather about it when they were alone.
The age of sixty-three was beyond his comprehension. It was
an overwhelming thought, that all of this man's life was behind
him, while all of the boy's was ahead. He never forgot that

moment, and many years later, when he himself was sixty-three and his grandson, John Maxim Lee was six, he opened the subject of age again. It was soon apparent that sixty-three was beyond this boy's comprehension too. The wheel had turned full circle. The little boy was now the grandfather who had his life behind him, and a new little boy had taken his place, with all of his life ahead. He summarized it philosophically, "Thus do generations of us men succeed one another."

Among young Percy's memories of that summer was a clear-cut impression of his grandmother, whom he described as "a very remarkable person." Her stature was short, even for those days, she being less than five feet tall. Her body, however, was vigorous and hardened. Her head was covered with gray ringlets, and the penetrating gaze of her sharp blue eyes reminded him of his sister Florence. She was a worthy daughter of old Brimstone Stevens, who was known among the Maxims for his ability to inspire both the fear of God, and of his own explosive self.

The summer passed pleasantly enough, instilling in Percy a lifelong love for rural life in Maine. When the day came for him to start home and get ready for the fall opening of school, there was a serious delay. Young Maxim couldn't get his shoes on. He hadn't worn them for two months, and his feet had grown. He pulled and tugged, and his grandmother pulled and tugged, but the struggle was hopeless. In typical Maxim panic, the grandmother shouted for Uncle Sam, who was waiting with the horse and buggy. He rushed into the kitchen, sized up the situation, and scooped up a cup of flour. This he poured into the shoes, shaking them about and dumping the surplus flour into a tin plate. The experiment worked, and Percy's feet slid into the shoes. But they were torturously tight, putting a severe damper on an otherwise delightful journey home.

That fall, about the time when he had achieved his tenth birthday, Percy sat in on his first business conference. The occasion was a family dinner to which Mr. Schuyler, one of Mr. Maxim's business associates, had been invited. The conversation was impressive, as these brilliant, inventive-minded men discussed the possibilities of the future. It was a time for a boy to be quiet and listen. It was also an occasion to remember, as Mr. Schuyler proclaimed, "Maxim, you may say what you like, but I can see the day coming when electricity will be generated in large electricity works and be distributed through the streets for house lighting, just as gas is generated in large gasworks today and distributed through the streets for house lighting."

Mr. Maxim did not agree with his guest. "No, Mr. Schuyler, you are looking too far ahead. Such a day may come, but there are too many unsolved technical problems for me to believe it will be in our time." This was in 1879, when the electrical arc lamp was not yet out of the experimental stage, it being too hot and too bright to be practical.

That dinner conversation spurred Hiram Stevens Maxim on to more intense creativity. The thing that impressed Percy most about this was that his father's experiments required the almost-constant services of a glassblower, and it was fascinating to watch him. Mr. Maxim was not so intrigued. There were many problems with the glassblower, and these were complicated by that gentleman's siege of whooping cough.

That winter, Percy kept pleading with his father to buy him a little play steam engine. Mr. Maxim finally acquiesced, after a wild scene in a toy store, with a thoroughly confused young lady clerk. The boy's mechanical passion was satisfied for the time being, but by the next summer he longed for bigger things. He wanted to ride behind a real steam engine. Since this seemed like a worthy ambition to Maxim Senior, he arranged for the boy to have a ride in the cab of a locomotive. This experience was not as pleasant as he had anticipated. The grease, dirt, noise, flames and hissing steam frightened Percy speechless. When pressed for his opinion on whether or not he had enjoyed the ride, he finally came out with an expressive, "Not very much." He still had a fascination for steam power, however, and the zenith of his young life was reached when his father presented him with a little steam train.

Percy continued his education in Brooklyn schools, and Maxim Senior went to England and to Europe on engineering business. Percy, who was twelve at this time, received a letter from his father, mailed from Paris and dated September 7, 1881: "It rains every five minutes here, and I lose lots of umbrellas." The missive continues with some international comparisons: "... Paris, which I think is humbug anyway. I am an American, and see things like an American, so I say Paris is all a hollow show, a complete sell. New York is the finest city in the whole world, and our own United States is the best country, and don't you forget it either." He goes on to say that Miss Haynes, his secretary, is going to write his biography "with a grand flourish," and "I will save an English copy for your Mamma."

It was at this time that William Thomas Hamilton was serving as Governor of Maryland. His teen-age daughter, delighted by life in the executive mansion, was Josephine Hamilton, one day to become the bride of Hiram Percy Maxim.

In 1884, at the age of 14, Percy began his studies at Massachusetts Institute of Technology. He enjoyed his courses in engineering, made many friends, and was such an able and entertaining speaker that he was chosen permanent toastmaster of his class.

When Percy was fifteen, and living with his mother and sisters in Hyde Park, outside Boston, he received another letter from his father, but not in his handwriting. In it he intimated that the family would all move to London, but this never came to pass, and father and son never met again.

In 1885 Maxim Senior received much publicity and acclaim for his invention of a flying machine. This was in England, ten years before the activities of the Wright Brothers. The March first 1885 edition of "The Implement and Machinery Review" featured an article describing this flying machine. It was forty feet long and fifty feet across, steam powered, with a crew of three. In an actual test it lifted itself 800 feet off the ground. It was not practicable, however, because of the tremendous weight of the water and boiler, and because of the danger from heat and steam.

In 1886 Percy was graduated from Massachusetts Institute of Technology as an engineer, when he was only 16 years old. It was a two year course then, but Hiram Percy Maxim was the youngest in his class, and in fact, MIT's youngest graduate. Later in the year he went to Ft. Wayne, Indiana, where he was employed at the Jenny Electric Company as an electrical engineer at the age of seventeen.

During 1887 and 1888 Percy worked as an engineer for the W. S. Hill Electric Company in Boston, nearer to his family and friends, and the familiar country which he enjoyed.

When Percy was nineteen, in 1888, he became engineering superintendent of the American Projectile Company, in Lynn, Massachusetts. This was a subsidiary of the Thompson Electric Company, which later became General Electric. A photograph of Percy taken at this time shows a perceptive, handsome face, dominated by large dark eyes. During these years he lived the life of a normal young man of his day. His activities included boating, fishing, bicycling, amateur theatricals, socials, and general mechanical experimentation. A flashlight photo made in 1890 shows Percy with three other young men having fun imitating a musical group, with shovel, saucepan, dipper and broom as "instruments." Percy is obviously enjoying this parody on the "swingers" of his day.

While Maxim, age twenty-three, was busy making projectiles for the Army and Navy, he also became involved with a project of his own, in the field of transportation. He had grown tired of walking, or riding a bicycle. Horses required a great deal of equipment, time and care. Street cars and trains did not always go just where or when one wanted to go. Maxim visualized something better, a horseless carriage. Because of the shortcomings in communication in those days, and because of the necessity for secrecy in races to the patent office, Maxim was unaware that others were working on the same idea. Those were to be the trial and error days of the Benz, the Stanley Steamer, the Daimler, and later of the Duryea brothers, of Haynes, Winton, the Appersons, and Henry Ford. In the beginning, none of these men were aware of the aspirations and experiments of the others.

For Hiram Percy Maxim, henceforth referred to as Maxim, or HPM, the horseless carriage era began late one summer night in 1892. He had spent a pleasant evening with an attractive girl in Salem, Massachusetts, and was pedaling his bicycle back to his home in Lynn. As his legs pumped up and down, his mind began to dream about a little engine, powered by gasoline, that could give the necessary thrust, so that a person might ride with more speed, and much greater ease. His thoughts were in the clouds, and through them he saw a vision. He pictured transportation emerging from its crude limitations. Faster modes of transportation would make towns closer together, and with motor cars there would be practically no limit to where one might go. People could intermingle, and this would affect occupations, marriages, and the very fundamentals of society and of civilization.

But the first need, in order to bring about this revolution in transportation, was a suitable engine. Maxim had heard of a little illuminating gas one that might have possibilities. It was known as the Otto Engine, and was used to run water pumps. It had a slide valve, and gas jet ignition. He thought that it might be adapted for mobility by using a hot tube. Its crank shaft was on top, and that would have to be changed. Also, gasoline would have to be substituted for the illuminating gas on which it operated. But it did run smoothly, and seemed to be the embryo engine of his dreams.

Young Maxim was unaware of the fact that in 1877, when he was still a little schoolboy, a man by the name of George B. Sheldon, living in Rochester, New York, had applied for a patent on the idea of a self-propelled vehicle, and that it had been granted, even though he had never produced a working model. Such a thing would not be possible, of course, under the patent laws of today.

Possibly the horseless carriage idea had come to most early inventors as they personally experienced the drawbacks of the crude transportation of their day. But HPM liked to think that it was an improvement born of the times, as one solution begat another. His theory was that when the bicycle became popular, it led men's minds to think of something still better. The answer had to be a motor car.

Once Maxim had settled upon the type of engine he planned to build, his next step was to learn about the properties of gas-

oline. But getting even enough of that volatile liquid to experiment with was a problem in those days. All HPM knew about
it to begin with was that it could be purchased in paint or drug
stores, that it would remove grease spots, and that it was very
temperamental. A great deal of work would have to be done
before he learned how to vaporize and ignite it.

On his way to work one day, he took an eight ounce bottle to a paint store in East Lynn and requested that it be filled
with gasoline. The proprietor was instantly and rightfully sus-
picious. HPM's dream was so wild that only youth could have
produced it.

After the factory closed at six o'clock, Maxim took out
his little bottle and studied it with a prophetic eye. In that col-
orless, mild-appearing liquid he foresaw a new world. In that
bottle he saw thousands of tiny drops, which, properly mixed
with air, could develop ten times the thrust of his leg power
against bicycle pedals. The contents of that little bottle could
take him all the way to Salem, if they were supplied with a
cylinder in which to explode them, and a connecting rod and a
piston to convert the explosions into mechanical motion. All
of these were compelling thoughts, but the inventor admitted
later that if he had realized all of the years of hard work, the
heartbreaking struggles and failures, the many new materials
and devices which had first to be invented, plus the thousands
of dollars that had to be spent, he would probably have tossed
the bottle onto the trash heap.

But fortunately all of these things were hidden in his future, so Maxim in youthful ignorance, proceeded with his crude
experimentations. He decided that what he needed first was a
simple get-acquainted test. To this end to took an empty six-
pounder cartridge case, readily available at the projectile
company, and which was two and a half inches in diameter and
twelve inches deep. He made a wooden stopper to plug the
open end. His idea was to get a drop of gasoline into this
chamber, then insert the plug and roll the case around a few
times so as to produce an explosive mixture. He then planned
on standing the brass cartridge case on end on the work bench,
removing the wooden plug and striking a match, which he would

quickly toss aside. He knew that this would not be a very exact experiment, but it would tell him a lot he needed to know.

Carefully he measured out one drop, and went through his planned routine. There was a brief and ominous quiet, and then, as he put it, "the end of the world." Fire shot out of the case, which reeled crazily on the bench. The match flew to the ceiling, and the young inventor realized that here was a thousand times more power than he had imagined. He knew too why the proprietor of the paint store had been so concerned. After Maxim recovered his equilibrium he repeated the experiment, using two drops, with about the same result. Then he tried three, and was able to detect a slight delay before the explosion, and also a little less violence. At length he reached a stage where the explosion was fairly dull, and resulted in a quantity of black smoke ... what we would call a rich mixture today.

But from that spectacular beginning, many more things had to come before mankind would have a motor vehicle. In that world of 1893 there were no spark plugs, no carburetors, no magnetos, and no dependable dry cells. There were no such things as clutches, gears, differentials, steering apparatus and pneumatic tires. There was not even an idea about where on a vehicle an engine could be mounted.

After months of basic experimentation, Maxim was ready to attempt a practical application of his idea. He bought a second hand Columbia tandem tricycle for thirty dollars. It was old, and its front wheel had a worn solid rubber tire only a half inch in diameter. The two rear wheels had one inch tires. Proudly the young inventor rode it to the factory, and stored it in a vacant room next to his office. He worked there alone every night until midnight. His first problem was to come up with a practical design. This was a large order, involving a chain drive, clutch, a gas tank and suitable support, an engine mounting, and most important of course, a good engine. All of his efforts on the designing board resulted in so many huge contraptions that he pictured them as necessitating an express train to support them, instead of one frail old trike. He was so discouraged with the results of his plans that he decided to build the engine first, and then figure out how to mount it. He later recognized this as an almost universal error among inventors. If one couldn't be successful with a pencil and paper, where the stroke of an eraser could correct a mistake, how could one hope to do any better with equipment already built?

The little engine Maxim designed as a three-cylinder, four-cycle air-cooled affair. The details of muffler, carburetor, manifold, and lubricating system he dismissed as minor details to be taken care of as they become necessary. This, he said later, was "a horrible example of how not to proceed." It took months of this after-hours work to finish the design of the engine, and more months to make the patterns, and get the casting and machine work done on the various parts.

When it was finally complete, Maxim surveyed his product and thought it was the most beautiful piece of machinery ever constructed. Carefully he set it in a frame where it could be cranked. Then he came head on with the carburetor question. He thought a great deal, and he read a great deal. Finally he hit upon what seemed like a workable idea. He took a small kerosene can which had a curved spout, soldered a copper tube in the bottom and led this to the needle valve of the intake manifold. When he opened the needle valve, gas dripped out. When the engine sucked, it drew in this gas. When it was not sucking, the gas ran down onto the floor. This operation, the young man found, was both simple and useless. It was also dangerous. Being a pioneer in the field, Maxim had had no experience in trying to start a cold engine. He spent an entire week in hopeless cranking. In so doing, he used up a great deal of gas. Sometimes he had to go to the paint shop twice a day to have his eight ounce bottle refilled.

After two weeks of this, the proprietor's suspicions turned to real alarm. He demanded to know what young Maxim was doing with so much gasoline, and the latter was afraid to tell him, for fear he would be arrested. The paint shop man glared over the top of his glasses and delivered a threatening sermon which concluded with the warning that "everybody who experimented with gas got killed doing it."

Maxim thought about the storeroom's gas-saturated floor, and nervously decided to attempt catching the dripping gas in some sort of container. This would save on the cost, and would reduce the fire hazard. Many years later, when he recalled this session with the paint shop owner, he thought how different that man's future might have been if he had had more vision. A great new industry had come very near him that day, and he had let it pass, unaware.

After months of cranking, the little gas engine still refused to start. Maxim decided to mount it on a lathe and use the shop power to turn it. Still no results. It was apparent that the mixture was too rich. Thinking the needle valve idea was no good, he tried using oil-soaked rags. This unscientific procedure failed too. Finally, late one afternoon, he decided to open the needle valve slowly, so as to get a gradual mixture. By this time the whole factory force was concerned with the problem of getting the little engine started, and they assembled at closing time, to watch. The lathe turned briefly, and then there was smoke, smell, fire and confusion everywhere. Some thing straight from Inferno had cut loose. Maxim's friend and helper, Leonard Stone, dived for the nearest door, with the rest of the men close behind. The inventor shut off his demon, and the sudden quiet was shocking. Blue smoke was everywhere. It was a discouraging moment, because although the engine had run, it was not a reasonable, manageable power, but an uncontrollable savage thing.

Maxim was not the type of person to suffer long from disappointment. He knew that the horrible noise could be controlled by a muffler. He would have to produce a positive-control mixture valve, and while hand cranking might be hard work, it was safer. He labored for a month before trying to start the little engine again. This time, after only fifteen minutes of effort, it started. It quivered, belched oil, and then stopped. Its inventor realized at once what had happened. Since there was no load on the engine, it had simply run away with itself. It had become too hot, and the pistons stuck. The spark coil drew so much current that it ran the batteries down. Maxim hadn't yet discovered that closing the throttle would control the speed.

With things progressed this far, Maxim couldn't wait any longer to mount the engine on the trike. Early one morning he scheduled his first run, without the prior precaution of a stationary test. He was young and impatient. Feeling both anxious and self-conscious, he mounted the vehicle and headed down the driveway. He had a foreboding that the unusual noise would attract a crowd, even at this early hour, and further, that disaster was brewing. Halfway down the driveway he became aware of it. He didn't have enough strength to pedal the contraption because the engine was connected by its sprockets and chain to the conventional driving mechanism of the trike. He had hoped that when the trike moved, it would turn the engine, but now he saw that this was impossible. It would have taken a team of work horses to achieve that. The obvious step was to take the chain from the engine and free the trike so he could push it to a steep hill. This he accomplished, with great effort.

At the top of the hill, Maxim's courage almost failed. It was incredibly steep. Its surface was rough, and liberally sprinkled with loose gravel. He gave a tremendous push and pedaled harder, but the engine did not start. Then he tried a leaner mixture. The results were immediate and terrifying. The trike lunged, careened, and slid. Gravel, fire and smoke were all around. Maxim was too busy trying to remain on the trike to be able to steer it, and manipulate its gadgets. He jerked on the handle bars as the front wheel plowed into a gulley, and the tire spun off and hung on the fork. The front of the contraption doubled up, and the back end catapulted over it. Dazed, Maxim got to his feet and took inventory. He had only minor cuts and bruises, and badly torn trousers, but his machine appeared to be a complete wreck. Gasoline was everywhere, fouling the air. Hurt in body and spirit, Maxim pushed the wreckage back to the factory.

It was apparent that a clutch was a necessity. Maxim worked on the idea during his spare time all that winter. When vacation time came, he planned to visit some other factories. The one which interested him the most was the Pope Manufac-

turing Company, of Hartford, Connecticut. He was acquainted
with the manager of its tube department, Hayden Eames. This
man, a graduate of the Annapolis Naval Academy, had been
the government inspector who checked on the projectiles at
Lynn. He was now retired from the Navy and occupied with
bicycle making. As a result of their meeting that day, Maxim
was hired as engineer by the Pope Company. He was to head
their new department of motor carriages. Maxim moved to
Hartford in July, 1895. And there, as he said, his "troubles
began."

The first plan was to take the gas engine off the little
trike and mount it on a Crawford Runabout, a four-wheeled
horse vehicle built along bicycle lines. Henry Souther, of the
Pope Company, assured Maxim that since the runabout required
only one horse to pull it, that he needed only a one-horse engine. HPM knew better, of course, and planned to discover
the exact amount of power needed by hitching up a horse and
inserting a pair of ice scales in the tugs. He did not reckon on
the objections of the lively horse, as the scales clanked noisily,
and banged against his legs. The power readings were too
contradictory to be interpreted.

In August of 1895 the 4-wheeled Crawford was ready for
a road test. Maxim didn't want an audience, so he and his
mechanic, Lobdell, pushed the carriage out of the Park Street
door of the factory at dawn. Almost immediately a crowd
gathered. This time young Maxim was not unduly excited. He
knew this machine would run. The question was how long, and
up how much of a grade. His aim was to achieve the three percent grade up Park at Zion, at a speed of fifteen miles an
hour. Eames was terribly excited, and bawled orders at all
the spectators in his best navy officer's manner. As the machine reached the top of the grade at the scheduled speed, ominous sounds began coming from beneath it. Maxim quickly
used the expiring spurts to turn the thing around, so that it
coasted neatly back into the factory yard. He and Lobdell
shook hands.

This machine was a real horseless carriage, right down
to the whip socket on the dash. It was noisy and smelly, and

completely revolting by modern standards. But Maxim had big plans for it. His next dream was a trip out of town and back, at least five or six miles. This was undertaken one night after dark, in order to avoid a crowd. Machinist Lobdell was to pedal along behind on a bicycle, bearing a supply of tools and repair items. Soon they had collected a following of about thirty other bike riders. They turned west on Farmington Avenue, down the worn macadam with the trolley track in the middle. Everything was going fine, and the young inventor was all but intoxicated with the sweetness of the engine's pull, until they came to the city line at Prospect Avenue. Then horrible sounds issued from beneath, as they passed the city limits, and the open country loomed dark and sinister. The strange sounds grew more threatening. Lobdell was relieved when Maxim decided to turn around before it was too late. Everything held together until they got back to the factory. Maxim's dream had materialized—he had driven into the country and back. This was Connecticut's first motor car, and Maxim was the first driver. The time was October, 1895.

A month later the Times-Herald announced the first horseless carriage race in America, to be held in Chicago on Thanksgiving Day. Maxim did not want to participate in it, as he knew that his machine had many faults yet to be corrected, but his employer insisted that it be entered. When Maxim learned that there were almost thirty other competitive vehicles, he was anxious to find out more about them, and how the other inventors had been solving their problems. As he said, "I had yet to learn that being entered for a motor car race is altogether different from participating in a motor car race, and altogether different from finishing in a motor car race."

Headquarters for the affair was at a large race track several miles south of the city center. It was an ideal site, as it afforded the participants a chance to try their machines, and to tinker with them. It took at least five hours of this per one hour of running. Half of the entrants had not even been able to get to the race track. Those who had made it were worried and dirty men, covered with the grease and grime of being "in and under" for several days and nights. Their vehicles comprised everything from a bicycle with a gas engine cylinder mounted on each side, to big clumsy wagons. Morris and Salom, storage battery exponents from Philadelphia, were there with a four-wheeled vehicle, and a backward-looking electric carriage with the steering wheel in the rear and drive wheels in front. These were called Electrobats. Charles and Frank Duryea from Springfield had the best-appearing entry—a regular buggy with a two-cylinder engine in the back. They were very confident of its ability, and tried to keep it covered from prying stares before the race. The German Benz was a strange contraption with a belt drive, which was allowed to slip to unclutch the engine. Both rosin and sand were necessary for its operation.

The night before this Thanksgiving race, a snowstorm moved in. The race had been scheduled for ten o'clock, and the drivers were in consternation. Since tire chains or snow tires had not been invented yet, there was a great rushing around to buy sash cord or rope to wind around the tires for traction.

Maxim was appointed as an umpire, assigned to the Electrobat. He kept pondering how this vehicle was going to negotiate fifty miles of wet snow five inches deep. Of all the entries, only six made it to the starting line. The Electrobat was one of them. It made its clumsy way down Michigan Avenue, stopping every few miles at the relays of storage batteries which had been deposited previously. It struggled into downtown Chicago as far as the Kimbal Carriage Company, when the battery died. That confirmed all of Maxim's feelings about the plausibility of using storage batteries to operate motor cars. The Duryea easily carried the field. The only other carriage to finish the course was the German Benz. Maxim learned more from this experience than he had from all of his previous undertakings. Building a practical motor car was not going to be the easy task he had assumed it was. Brave men with lots of tools, knowledge and spare parts, and who were unafraid of dirt and grease and noise were necessary before people were going to be speeding along over the countryside.

It followed that Maxim's next assignment was to design and build an electric horseless carriage. It was to be called the Mark Number I. The reasoning was that electric power for a vehicle would be clean, quiet and reliable. Even though it had serious drawbacks, it could fill in until ingenious men could eliminate the noise, grease and frequent breakdowns of gasoline motor cars. The Pope Manufacturing Company was a leader in the field of bicycle production, so it was important that this new venture should prove successful, and thus enhance their fine reputation. The responsibility was Maxim's.

It was decided, in the interests of publicity, to give a public demonstration of the Mark I, and Maxim was assigned to do this at eight o'clock one evening, on Weathersfield Avenue, in the city of Hartford. In preparation, he had ridden his bicycle over the route and checked on grades, holes, and crosswalks. He made arrangements, unknown to his employer, to have mechanic Lobdell follow at a respectful distance in a horse-drawn express wagon. This vehicle was loaded with storage batteries, and spare parts.

Maxim set out that night, as he said, "nervous as an opera singer." It did not add to his confidence when he learned that he was to have a passenger on the run, and that it would be Mrs. Day. She was the wife of the vice-president of the company. Bravely Maxim started the vehicle and wheeled around the block eight times, with Lobdell's wagon tearing along behind him. This made quite a spectacle for the large crowd which had gathered. Bicyclists, pedestrians, drivers of horse-drawn vehicles and people in trolley cars all watched and cheered this history-making demonstration.

During his motor car experimentation, Maxim kept careful records in neatly indexed, leather-bound workbooks. They were stamped "Western Electric and Manufacturing Company, Pittsburg, Pennsylvania," and were all in his own handwriting, in pencil, to allow for corrections. These notebooks included scale maps of the routes used for testing, and the mileage records of the various models. They were identified by number and location, such as the Harlem Route Test, the Fulton Market Test, and so on. A typical entry reads: "Mark I memo

.... for 10 mile speed, fall in voltage of present #2 batteries makes it impossible to calculate what induction is necessary to get torque and speed to correspond. Mark I weight, 1550 pounds (with passengers, 1825), sample test run... mileage 9.75, time 55 minutes... 10.63 miles per hour, stops included."

After the success of the Mark I, Maxim developed the Mark II. It was a two-cycle job, and after a few miserable runs around the factory yard, it turned out to be the engineering failure of his career. He was summoned to the company's main office, and told that Mr. Pope himself wanted to see him. Fearful of losing his job, Maxim went to the interview, which also included Eames, and vice-president Day. But instead of being fired, Maxim was assigned the project of building a small gas trike, suitable for merchants to use in making package deliveries. It was to be known as the Mark VII. At the same time, he was to be working on the design for an electric phaeton to be called the Mark III.

In February of 1897 the Mark VII delivery trike was ready for testing. Maxim waited patiently for the spring thaw, so that he could take his vehicle out of town for a trial run. His objective was a twenty-five mile trip, to Springfield, Massachusetts.

Accompanied by mechanic Lobdell, Maxim set out one night, with the aid of a kerosene bike headlight. As soon as they were off the macadam, the road was nothing but mud. The little machine wallowed and slid, and its cylinder head glowed red in the darkness. Lobdell peered ahead, apprehensive of twenty-five miles of such conditions, but Maxim did not want to give up, because the engine was still running. After some indecision, they tried following the trolley line. This was terribly bumpy, and the ground between the ties was too soft. They got as far as Hencoop Bridge and gave up. They had gone less than a mile beyond the city limits of Hartford.

They waited for two weeks for the road to dry, and then prepared to try again. They had improved the engine by adding a hot tube lamp, and new and better bearings. They had also learned more about operating technique. Hopefully they set out, with Lobdell sitting on the tool box lid, and Maxim in the saddle. The road was better, and they chugged along, scaring numerous horses. Then the terrain got worse, and the jolting became serious. Since the motor was constructed to run wide open, trying to go slowly was difficult. Large holes and rocks wrenched the little machine horribly. Suddenly, with a loud crack, it raced at terrific speed, literally tearing itself apart. When it suddenly stopped, the riders found that the drive wheel had come loose from the axle. Since this was a serious breakdown which they could not possibly fix on the road, Lobdell and Maxim walked to the nearest farmhouse. By coincidence it happened to be the home of one of Connecticut's most distinguished inventors, Mr. Christopher Spencer. The lady who answered the door was already familiar with the problems of inventors. She gave them permission to push their gas trike into a shed, and they walked back to the city line and caught a trolley for home.

For the next two days they carried tools and parts out into the country, secretly, so that no one would know of their failure. Since the trouble was that a key in the driving axle had sheared off, they learned the important fact that something else had to be developed. Maxim's solution was the multiple-splined shaft, one of the most important features of automotive engineering.

In May of 1896 Maxim planned his third trial. By this time the roads were much better. They made it into Windsor and beyond, affording many people along the way their first view of a motor vehicle. Rolling along at from 10 to 12 miles an hour, they found it necessary to stop for every horse they met and coax it past. The thrill of touring over the lovely countryside was further punctuated every fifteen minutes by a stop to pour a cup of heavy cylinder oil into the crank case. But by the standards of their horse-and-buggy age, they were achieving real speed.

As they approached a railroad crossing along the river,

they encountered a horse and carriage. The animal was extremely upset, and as Maxim descended from his vehicle to help, the driver leaped from the buggy and took out across an open field, deserting his frantic passenger. Maxim recognized the man as a neighbor of his from Hartford, and smiled to himself as he took the horse by the bridle.

After leaving Windsor they found themselves to be "the biggest sensation since the hotel burned." They struggled along in the dark, with the aid of their kerosene lamp. The oiling, the chuck holes, the adjusting, and the horses, had cost them a great deal of time. Hurrying to make up for this, they came abruptly upon something looming out of the darkness. It was a horse-drawn junk wagon, with a terrified driver. His frightened animal cramped the vehicle and buckled a wheel. The petrified man, who had never seen a motor car before, even in broad daylight, was unable to tell his name, or where he was going. Maxim did the best wagon repair job he could with wire, and sent the outfit on its way.

It was late when they finally saw the lights of Springfield. A startled policeman directed them to the hotel stable yard, where a man dashed out of a shed, waving a lantern. Exhausted as he was, Lobdell couldn't resist this opportunity to have a little fun. He thrust his face close to the old man's and asked excitedly, "Is this Philadelphia?"

Bewildered, his victim replied, "My God, man... no! This is Springfield, Massachusetts!"

Lobdell and Maxim shook their heads solemnly. "He says this is Springfield."

Later the stableman said he was so dazed by this startling commotion at three o'clock in the morning that he thought one of the Boston and Albany locomotives had jumped the track into his stable yard.

It had taken Maxim nine and a half hours to drive twenty-five miles, but this was the first time that a Connecticut machine had been driven out of the state under its own power.

Maxim had also been continuing with his assignment for the design of an electric car. He wanted to use a great deal of nickel plate, but his employer said that he wanted "a gentleman's carriage," not a fire engine. In May of 1897, ten of the vehicles were completed, and the company held a combination celebration and showing. Tours were taken through the factory, refreshments were served, and special guests were treated to rides. One of the notables was General Nelson A. Miles, the frontier officer after whom Miles City, Montana was named. When a publicity photo was made of the Mark III, some feminine charm was added. The patent department selected what they termed "the smartest looking young women in Hartford" to ride in the carriage. The sales department had a real challenge, because these electric vehicles sold for three thousand dollars.

Later that summer a social gathering was scheduled by the Eames family, in a charming house on a Farmington hill. Eames decided to ride out to the affair with Maxim on the Mark VII. The road was too soft for the larger and heavier electric machine, and the distance of sixteen miles was too great for its storage battery power supply. Maxim fastened two office chairs onto the little gas trike. There wasn't much to hang onto. They overtook and passed the trolley car, and waved proudly to their friends inside, who were also heading for the Eames place.

At dinner that night, all the talk was of the successful trip of the gas-powered machine. Recklessly jubilant with such success, Maxim offered to give a ride to any of the young ladies present. Mrs. Eames politely and firmly declined, but her sister Julia accepted.

As he made his preparations, young Maxim began to worry. The Eames house stood at the top of a steep hill, and the Mark VII had no brakes at all. The road led down the incline, and onto the highway. Justifiably, Maxim was fearful of the turn at the bottom.

Miss Hamilton was a little frightened too, but she was a charming good sport, and climbed gingerly into the rear chair. She was overcome with the desire to be able to say that she had ridden in a horseless carriage. Lieutenant Eames boarded too. The engine exploded into a start, and Maxim was busy trying to use the clutch as a brake, and keep the machine

in the softest part of the road. He kept thinking about the time
he had been catapulted over the front of such a vehicle in Lynn,
and dreaded having this happen to his boss, and to this lovely
young lady. Exercising all of his wits, he safely negotiated the
turn at the foot of the hill. The engine performed perfectly,
and the ride was thrilling, with Miss Hamilton showering hairpins all along the route. They toured the village, and made it
back up the hill.

The whole affair was proclaimed a success, and Maxim
left for home that night rejoicing. In his visionary mind his
plodding, noisy machine became swift and silent. Coupled
with the pleasant fact that at this late hour all of the horses
had gone to bed, it was a moment to savor.

The year of 1897 was important for another Maxim also.
Uncle Hudson developed smokeless powder. He eventually sold
the idea to the DuPont Company, and was employed with them
as a consulting engineer.

For a year, beginning with January of 1898, Maxim kept
a diary. He wrote in fine neat script in a little three-by-five
inch notebook, whose leaves were headed "Columbia Bicycles,
Pope Manufacturing Company, Hartford, Connecticut." A
January entry reads: "Went over to the Canoe Club to see To-
lez. Took mother. Scared horses in front of Linden, and made
a bad fall in the street."

On February 26 he wrote: "Dictated gasoline vehicle report. Feel rotten mean yet. Bad cold. Submitted design for
new flat solid tire to Coles. Had the most ridiculous conference probably ever held by educated and intelligent men."

On March 8th he bowled at the Casino with his sister
Florence. March was a month of double destiny. The entry for
the 16th states: "Spent all day thinking about noise and shaking." (Foreshadowing his invention of the silencer.) "Lunched
at Eames. Conference on countershaft for Mark VI. Met Josephine Hamilton. Impressed me as older than I, and very
conventional."

Evidently Josephine did not share her sister Julia's enthusiasm for crude motor cars, for the April 3rd entry reads:
"Wonder why the fair Miss Josephine Hamilton didn't want to

go out in the Mark IV. Guess it is too plebian."

Three days later he records that he "took the dignified young lady home. Actually induced her to smile once. She may be mortal after all."

On April 11th the notations introduced a new factor, the trouble with Spain over Cuba. He wrote: "Miss J.H. certainly is... well, we spent the entire evening together." He goes on to say that the main discussion topic was the threat of war, intensified by the news that all the Americans had left Cuba.

April 14th Miss Hamilton left on a trip, and Maxim saw her off on the train. That night he wrote about "her blue eyes, and her blue Easter hat."

On April 19th his diary states (about the war): "All hope abandoned now. War is inevitable in all probability." And again on the 22nd, "Everybody is disturbed over war. War practically begun today. Spanish merchantman captured by our Nashville." But life in Hartford apparently went on, because he concluded with, "Card party in the evening."

By April 25th Maxim had begun to consider his own military involvement: "First talk at home about my having to go. Mother says she couldn't stand it. Thinking about plot for a play... the plot thickens... the Josephine Hamilton plot."

On May 2nd, he received a letter: "Which brings me nearer the verge of the precipice than I expected to be in many years. I wonder if I have at last come to it? Nonsense!"

Two days later his thoughts returned to the factory drafting board: "Thought of a bicycle ratchet clutch essential. Went downtown to see soldiers off."

Apparently Maxim's ideas were becoming clear to the family, for on May 8th he wrote: "Mother said she did not think it ever advisable to live with young married people."

During this time Maxim was absorbed in some amateur theatricals, which he always enjoyed. A group of young people was putting on a hilarious parody version of "The Rivals," and he was playing Lydia.

May 18th presented a problem faced by many young people when they make the big announcement: "Terrible scene at the house this noon. Mother and Florence in tears over J.H."

But the car tests continued on schedule. "Entz came. Took him over mileage run number 38... 17 miles."

On May 21st Maxim took an important walk: "to 210 West Washington Street (Hagerstown, Maryland), with roses, and got first glimpse of Hamilton house. Doggone nervous."

A scrapbook picture dated May, 1898 shows Josephine smiling provocatively from behind a lacy parasol, standing in a wooded area beside a large three-forked tree. Her hair is curling softly, enhancing her large, dark beautiful eyes. A group picture of this period includes Hayden Eames, distinguished appearing, with dark hair and a moustache; Mrs. Eames, a large pleasant-looking woman; and young HPM, natty in white ducks. The album perpetuates many memories of Maxim's courting days .... with J. H. in rowboat, sitting with J. H. under a tree, standing on a log .... J. H. attractive in a white blouse with a high collar and a brooch at her throat, combined with a dark gored skirt and a wide belt. In all these pictures, the word for Maxim's expression is "attentive."

To continue with his diary: "May 26th, suggested to Pope opposite revolving case and vane rotary hydraulic motor a la Dewey. Electric balance gear. Went to Buffalo Bill show in pouring rain."

There is no clue to the single entry on May 29th: "Wrote to Maude Adams." She was a famous stage actress of that day.

By May 31st, things were serious again: "Letter from Josephine about ancestors. Extras about fight in Santiago."

In June, Maxim was interested in pictures. He received ten from Josephine in the mail, and went out and bought a new roll of film for his own camera. He wrote: "Cost four dollars. Gave Florence a five dollar gold piece for her birthday."

June 10th was a big day: "Bought engagement ring for Josephine at Tiffany's." There is no explanation for an entry several days later which refers to "that damned ring."

By August, things had gotten down to practicalities. Maxim wrote on the 27th: "Josephine says she will not in December unless I can do it from what I save from now until then. Going to try and get up something."

On August 29th appears the statement: "Gave Josephine

my horoscope."

This handwritten document from Professor St. Leon of New York is today part of the Maxim Collection in the Connecticut State Library at Hartford. Whether or not one believes in the validity of astrology, this reading was remarkably accurate. HPM was a Virgo, having been born on September 2, 1869, Thursday at 9:30 A.M. The reading includes an accurate health comment: "Strong constitution, good stamina, physical strength and activity." It mentions risk of serious accidents, but few materialized. It continues: "The planets show a great liability to take cold in the head or throat." This proclivity bothered Maxim all his life, and eventually led to his death. The horoscope told of a fortunate courtship, and an early marriage which would be happy and comfortable. The wife would have brown hair and be of a good disposition, and of a good family. There was to be a probability of only one union, and the family prospects of many children. Actually, there were to be only two. Many changes of residence were indicated, and one or more long voyages. All of this came to pass.

The reading indicates that Maxim had "good intellectual and artistic powers.... literary and scientific tastes, curious and original ideas, and peculiar opinions." It states that he was to be "ambitious, determined and bold, fond of pleasure and argument." It indicates "evidence of prosperity and success, but life will not pass without many bothers and complexities, and perhaps some litigation." This too was proved accurate. The reading indicates "no trouble shown with relatives or neighbors," and no need to fear enemies, because there would be few, if any."

But the thing about this horoscope which appears so remarkable today is that it indicates success and fame in a field which at that time was unheard of, namely, amateur radio. It mentions "much honor and good fortune ... aid from influential people, and much popularity." It flounders a bit with the idea of acting as a profession, but goes on to say, "Any occupation that brings contact with the multitude is favorable." What a comment about the man who was to become "the father of amateur radio!" The document concludes: "This is a very fortunate horoscope.... the planets show the possibility of a brilliant career and a happy destiny. Be it yours to serve them. With thanks for your confidence in the sublime science of astrology........"

Maxim's 1898 diary continues with July 31st: "Had to push trolley car from Electric Park." In those days, men passengers frequently had to get out and render assistance when public vehicles stalled. The day concluded on a happy note: "Took first photos of Josephine."

On August 7th Maxim went canoeing with Josephine in the morning, and recorded that he hoped she would "be entirely unconventional someday." In the afternoon he resumed his motor car testing, and wrote that there were "problems with gears, clutch, transmission, flywheel, generator, rheostat, roller crank, and piston pin bearings." Many years were to pass before all of these problems would be solved satisfactorily, and before the public mind would be ready to accept motor cars and motor trucks on a general scale. Convincing people of their reliability was only half the problem. There was also the difficulty of getting potential customers adjusted to increased transportation expenditures. The early Columbia electric carriages cost three thousand dollars, a real fortune in those days. People laughed at the very idea, saying that they could get a good horse and buggy for four hundred. Even inventor Maxim himself was cautious, saying, "I was the most optimistic of the optimists... but even I never dreamed motor cabs, motor trucks, motor buses, and private motor cars would be used in the millions."

Maxim's August 10th diary entry reads that he "sat on the Inn piazza all evening with J... Talked plainly about morals of the average masses. Thank God I am clean and able to look her in the eye."

On August 13th he mentions a walk "toward New Britain with J. Little kitten followed all the way."

September 2nd of 1898 was HPM's 29th birthday. He made a test car run: "Started trip to Boston at 7:00 P.M. Got to Thompsonville soaking wet."

On September 8th the diary reads: "J. sent out formal announcements of our engagement today." Six days later there is this entry: "Letter from Beatrice asking me to come. Very startling, to say the least."

Beatrice was a former girl friend. Maxim had named his first boat after her, and kept her name painted on the craft for years.

The diary was casual: "Pope gone to shooting match in Scranton. Wrote Beatrice telling her of engagement."

On September 18th Maxim was ready for a trial run of the Mark VIII. This vehicle featured a bonnet (hood over the engine), and rear axle drive, as of today. It had a foot feed carburetor and a radiator in front. It was the first horseless carriage to look a little like a modern motor car.

Mr. Hart O. Berg, "a fastidious gentleman" who was the Pope Company's European representative, was assigned to go with Maxim on the test trip. HPM was worried about how this man was going to endure the roughness of the road, and the rearing horses. It would be a long, dirty ride, with the possibility of rain, from which, of course, there was then no protection. The route was to be the Boston Post Road, which was in such a miserable condition that even a modern motor car could not have achieved a speed of fifteen miles an hour on it, if indeed it could have negotiated it at all. The way was one of rocks, ruts, and weeds. There wasn't a filling station between Hartford and New York. In fact, there wasn't one between Hartford and San Francisco. The only service available was in blacksmith shops. The only gasoline was in paint or drug stores.

Since there were no road maps, and no road signs, the two men got lost. To add to their problems, the law required that when the driver of a horse held up his hand, the motorist must pull over to the side of the road and stop until the horse had been led past. The gentlemanly Mr. Berg did not relish this latter assignment. As the day fagged on, he became quite profane, in French, for greater expressiveness.

Beyond Wallingford they encountered a Sahara of sand. They had to stop frequently to cool the engine. When they finally achieved Stamford about dark, Mr. Berg decided that he had a pressing engagement in New York, so he boarded a train, leaving Maxim to carry on alone.

The young inventor became lost again, in the darkness. It was very late when he finally found the bridge across the Harlem River, and for years afterwards, he equated Harlem with Heaven. There was no garage in all of New York, and no livery stable would take him in. Fortunately there was a night watchman on duty at the Pope Company's New York headquarters, in what was later to become Columbus Circle. After a lengthy cross-examination, Maxim was permitted to drive his machine inside. The next day Mr. Berg, surprised that HPM had made it, agreed to undertake the return journey.

Going up a rocky hill out of Rye, they heard a terrible crack, and a thud. Maxim alighted, and surveyed the ruin. The driving gears had fallen out, and lay hopelessly in a pool of oil. The men had to finish the trip by train, after arranging for the carriage to be returned by freight. The result of this experimental journey was the conclusion that the engine and gears would have to be mounted so that they would not drop out on bumpy roads.

A company argument ensued about engine design. Maxim wanted two cylinders, but the officers insisted that one was enough, and that anything more would be too complicated. Their Mr. Atwood detested gas engines. He couldn't foresee, in 1898, that they had any future whatsoever, and he resented the company's spending money to develop the messy contraptions. Mr. Day also felt that gas machines were too dirty to have any success with the general public. He studied a set of filthy cog wheels in the model room one day and demanded to know if they were necessary. When Maxim assured him that they were, he shook his head and walked away, saying, "Let me tell you something. We are on the wrong track. No one will buy a carriage that has to have all that greasy machinery in it."

Mr. Day was one of the leading minds of Hartford, and his views were respected. But Maxim refused to give up, and never gave the order for the work to stop. Years later, when

Day became president of the Association of Licensed Automobile Manufacturers, Maxim chanced to be walking with him up 5th Avenue in New York. He gestured toward the torrent of traffic and reminded his companion of his old remark about the improbable future of gas-engined cars.

Mr. Day's reply was humble: "Great Scott, Maxim, do you remember that? I was awfully wrong, wasn't I?"

It was about this time that Lieutenant Eames returned from a trip to Europe and reported that the French had solved the gear problem by "driving with their feet." He proclaimed that such monstrosities would never be adopted by the public, and that they would baffle the abilities of a locomotive engineer or a steam shovel operator.

During this year of 1898, Eustace B. Entz, chief engineer of the Electric Storage Battery Company of Pennsylvania, became involved with an idea. He proposed to create an electric car with a simple control lever for all the gears. It had the drawbacks of great cost and weight, but it was still a clever idea. On September 24, Entz, Pope, Maxim, and a mechanic took this machine for a test run. It was a complexity of wires, voltmeters, ammeters, and other electrical devices. Entz drove; Maxim kept the engine running; Pope read the instruments, and the mechanic was assigned to do the required fixing.

The car started smoothly, but balked on the short upgrade because of the weight of the men and all their equipment. The gas tank was a copper box fastened to the dashboard. An ammeter with a naked terminal lay on the floor nearby. Maxim jumped out to make an adjustment, and his foot caught in one of the tangle of wires, so that the terminal hit the copper gas tank. The resultant arc melted a hole in that container. The gas ignited, and in seconds the automotive pioneers had a roaring fire in the middle of Laurel Street. Gas ran all over the carriage and onto the ground. Pope called the Fire Department, but before they arrived with an impressive clatter and clanging, a small boy had extinguished the blaze with a garden hose.

Maxim had sprained his ankle in the escaping leap, and

it pained him severely that night as he got into a hansom cab to escort Josephine to the theater. Entz did not recover so quickly. His nerves had suffered almost as much as had his machine. However, he never lost faith in his idea. He took it to the White Motor Company of Cleveland, where it was developed as the Owen Magnetic. It was never destined for success.

The beginning of the end had come for the horseless carriage days. This was hastened by a discovery made at this time by Herman Cuntz, of the patent department of the Pope Manufacturing Company. This gentleman found out that a patent had been issued November 5, 1895 to one George G. Selden of Rochester, New York, on the idea for a self-propelled carriage. Mr. Selden had applied for this patent in 1877, and it covered any vehicle powered by a hydrocarbon engine with a disconnecting device between the driving engine and the vehicle, and a receptacle for liquid fuel.

The discovery of the existence of this patent spelled trouble for the Pope Company, because the hydrocarbon engine was its gas engine, the disconnecting device was their clutch, and the receptacle their gas tank. It was impossible to eliminate any of these features, and Mr. Cuntz was justifiably worried. It was apparent to his trained mind that this Selden Patent gave its owner definite rights to a monopoly on the manufacture and sale of any gas-powered vehicle. Since it was Cuntz's job to warn chief engineer Maxim against infringement, he did so in plain and dire terms, ending with the insistence that all motor carriage work be stopped immediately. Maxim was derisive, explaining that the engine described in the patent was completely impracticable, had never been built, and was little more than a joke. Mr. Cuntz replied that it was the wording, not the drawings, which made the Selden Patent a very real threat. He pointed out that a United States Patent was valid until the courts ruled otherwise, and that any infringement could make the violator liable for a suit.

Cuntz's next step was to take up the matter with Hayden Eames and Mr. Day. Neither of them knew what to do. Cuntz's information was just too terrible to believe. It meant the termination of their prospering young business. Helpless, they did nothing at all about the matter. Cuntz was shocked at their attitude.

Meanwhile, the Pope Manufacturing Company began negotiations for a deal with the Electric Storage Battery Company in New York, to organize a cab-operating firm. Mr. Whitney, former Secretary of the Navy during the Cleveland administration, was a leader in this project. His son, Harry Paine Whitney, had bought a Mark VIII gas carriage, one of the first out of the factory. It was shipped to New York, where it was scheduled to be used on runs to and from his country estate on Long Island. Since fashionable New Yorkers were beginning to manifest an interest in motor cars, it was very important that Mr. Whitney's carriage prove altogether satisfactory. Therefore it was a dreadful development when HPM was summoned to Eames's office one day and confronted with the statement that Mr. Whitney's Mark VIII had "failed utterly."

Maxim was aghast. His tongue, true to a habit when he was confronted with bad news, tasted horribly of copper. This "utter failure" reflected accusingly on HPM, since he was the chief engineer, and the one responsible for the vehicle's design. Anxiously he inquired if anything had broken. It had long been his worry that an axle or steering gear might snap and injure somebody. But Eames explained that the failure was in the engine. It refused to start, even for the experts. It was therefore Maxim's job to go to New York and start that car. This assignment was made with the hint that if the company's chief engineer was going to have to go about the country starting cars, then it was time some changes were made.

Baffled and worried, Maxim went to New York, where he found the offending carriage in the loft of a street car barn on Fourth Avenue. It appeared to be clean and shiny, and well cared for. It was apparent to him at once that something strange was behind the problem. There was nothing wrong mechanically. The compression was excellent, but there was an abnormal drag. Maxim's experienced judgment told him that this was not due to a tight bearing, or a scored cylinder, or a dragging clutch. What could possibly make an engine act like this? The inventor tested the oil, and found it all right. There was a satisfactory spark. Finally Maxim decided to check the gasoline. Jouncing the car assured him that there was liquid in the tank. A check on the carburetor proved it to be intact, but when he jiggled the float valve, he got the wrong reaction. It was sluggish, and wouldn't flood. When a little of the liquid finally oozed through, Maxim examined it closely. It was extremely sticky, and had a familiar odor, which Maxim recognized as that of varnish. He took the carburetor apart and found it such a gummed-up mess that it was impossible for any part of it to function. A test of the fluid showed that it would not ignite. The cause of the "utter failure" was all too clear. The gas tank was full of varnish. Maxim pondered his next step. There was no hope of cleaning out the stuck areas. The entire machine would have to be shipped back to Hartford, so that all the affected parts could be replaced.

Pointedly, Maxim asked Mr. Whitney to send for the man responsible for looking after the carriage. Meanwhile, the inventor went to a drug store, where he purchased a small bottle and a cork. He then re-assembled the feed line and other parts, to give the man, when he should come, the impression that Maxim too had just arrived. HPM led the man into talking, and he quickly made known his stand. The design of the engine was all wrong, he maintained. It was apparent just to look at it that it would never work. The cylinder was too long, and the valves were wrong. The spark plugs, carburetor and tank were all in the wrong places, making feeding impossible. Maxim recognized this as a plot to discredit his machine, "good old-fashioned political manipulation such as we have been cursed with for centuries," he termed it. In full view of the man, after looking him straight in the eye, Maxim took the little bottle from his pocket and proceeded to fill it with a sample of the fluid from the gas tank. He corked it tightly and placed it carefully in his pocket. Then he informed the man that he had been a witness to the action, and that anyone with his experience should know why that engine had failed to start.

The next day Maxim reported back to Lieutenant Eames.

"Well, Maxim," that gentleman demanded, "did you fix Mr. Whitney's car?"

"No, sir," was the flat response.

Maxim was reluctant to tell his story. He was despondent over this sort of dirty work, and was concerned about how far up the line such trickery might extend.

When Eames showed his irritation, Maxim admitted that he knew what the trouble was. He showed his boss the sample of varnish, and related the entire story. Whereupon a new carriage was shipped to Mr. Whitney, with the suggestion that in the future he see to it that something more nearly like gasoline be put in the fuel tank.

Before long, there was more trouble. Reports came in of a new carburetor disease. Engines passed their factory tests, and then failed completely in service. A team of New York experts worked for days vainly trying to come up with a solution. Pope Company production had to be held back. It seemed that there must be something very wrong with an engine if only the chief engineer could make it run. Maxim was sent to investigate, and diagnosed the trouble as gas stoppage. Something rattled in the carburetor, and it shouldn't have. Maxim ordered a hole drilled in it, and found a little copper rivet. This had been serving as a valve. With a gentle flow of gas, the rivet would settle to the bottom. When there was a sudden, strong flow, the rivet would rise and cut it off. It was a clever piece of sabotage.

Maxim plugged the hole, reassembled the carburetor, and started the engine. He raced the carriage out of the factory in front of the office of Mr. Budlong, the company president. Later he presented the gentleman with the offending rivet, and a detailed explanation.

Some weeks later, Maxim was again summoned to the test shed. They had an engine that no one was strong enough to start. Maxim examined it and found that the bearings were loose enough, the mainbearing free at the flywheel, and finally that the shaft too was loose. All this, and yet the engine was so stiff that it could not be cranked. Naturally, Maxim smelled another rat. When he got a test lamp and investigated the inside of the engine, his trained eye located a tiny bit of oily rag. Obviously such a thing had no business being in the crank case of an engine. He poked about with a screw driver and came upon a great mass of something soft wedged between the end of the crank and the crank case. Not letting on that he had discovered anything, he called the foreman and asked him what he thought the trouble was. The man promptly rose to the bait, and launched into a long speech on the many faults of the engine's design. He said that there was absolutely no way to keep the bearings in line, and that if the company was interested in producing a marketable machine, its first step was to junk this model.

Maxim listened quietly, wiping the grease from his hands. Then he turned to the foreman and said casually, "Take that wad of rag out of the crank case."

The unhappy fellow poked around in seeming ignorance until another worker stepped forward to help him. Together they pulled out the unravelings of an entire shirt! There was nothing more to be said, since the plant superintendent had witnessed the whole affair.

In October of 1898 Maxim's wedding plans were advancing, but not always smoothly. His diary entry for the 31st records, "Florence declined to be J.'s bridesmaid because she must wear pink."

By November Maxim had reached the point where he did not wish to share Josephine's company with anyone else. On Sunday the 20th he wrote, "J. left me alone in the evening to see callers. Terrible blow."

On November 24th he had an important motor carriage meeting with some German representatives, but he still was concerned enough to write, "No letter from J. again." On the final two days of the month, he went shopping: "Went with Mother and selected silverware. Ordered new frock coat. Bought new hat. Paid Barton twelve dollars for dog. Worked on engine tracing."

On December 2nd the entry reads, "Bought the wedding ring today... happy day." On the 8th he records that he "had to use hot water to get the trike started"; on the 9th, "article in Courant about wedding." On the 11th he was in Hagerstown and saw the wedding presents. He notes that he was suffering from "a bad cold." On the 15th Maxim wrote, "Had salary raise notification."

On December 21, 1898, Hiram Percy Maxim and Josephine Hamilton were married. That Christmas eve they sailed from Hoboken, New Jersey, on what was known then as The Grand Tour. This lasted for weeks, and included visits in Holland, Belgium, France, Germany and England.

On February 7, 1899, while still on this honeymoon tour, HPM wrote a letter to his mother-in-law, Mrs. William Hamilton. It was in dark ink, beautifully shaded, and postmarked from London: "I am banking upon being counted in the family now," he stated. He went on to compliment his wife, Josephine. "We have traveled all day long in an infernal little railway carriage compartment with a third person whose presence, and talk for hours in succession, drove even the paint on the walls to the confines of revolt.... but Josephine would make the most querellous crank on this footstool a paragon of sweet temper and easy disposition." The family album includes many pictures of this tour, including the sea and the ship.

When the Maxims returned from their honeymoon trip they settled in an apartment on Capitol Avenue in Hartford. This was in a brick hotel with a sort of turret on top, not at all the type of domicile to which Mrs. Maxim had been accustomed. Interior photographs show a small mantelpiece with an embroidered cloth, and floral wallpaper of large design. It was located past the State House and Constitution Hall, and past Asylum Street, named after the deaf and dumb who resided there. HPM walked it many times, on his way to and from work. On August first they moved to Park Terrace, in the corner one of a long row of two-and-one-half story houses. There is a picture of this place with Josephine standing on the front steps. Other photographs of this period include one of Maxim and Eames in the Farmington Golf Series, and one of the first gasoline autos made in Hartford, taken on Capitol Avenue, in front of the Pope Manufacturing Company buildings.

One of the first men in Hartford to purchase a horseless

carriage was Leonard Fisk. His brother owned a half-mile race track in Branford, Connecticut, and he conceived the idea of holding a motor car race there. This was the first such race in America. The year was 1899. Maxim was determined to get into that race with the Mark VIII, but when he approached his boss, Lieutenant Eames on the subject, he was rebuffed. Eames reminded him that the eyes of the whole country would be on that race, and if the Pope Company was to have an entry, it would have to win, or the reputation of the firm would suffer.

Maxim was undaunted, and arranged for his model room foreman, Fred Law, to be at the factory at five o'clock in the morning to commence the drive to Branford to enter the race. He wanted to get there early, so that he could look over his competition. If they looked sufficiently inferior, he would wire Eames that he would enter, unless ordered to the contrary. If the other machines looked too good, he would be satisfied with being just a spectator at this historic event.

Although they left Hartford at five A.M., and the race wasn't scheduled to start until two P.M., the resulting nine hours was not enough for them to drive the forty-two miles! At Meriden the chassis frame broke, and the gear box crashed down onto the road. Since nothing but a blacksmith shop could service them, the men walked back to one they remembered passing. They routed the owner out of bed, and got him to do some forging. When the machine was running again, Maxim contacted Eames, who gave in to the point of leaving the matter entirely up to HPM. That was all the inventor needed, and he pushed on. With only three hours to cover the remaining twenty-four miles, he encountered a wilderness of sand. His engine boiled over, and the only available water was a little pond covered with green slime. The men skimmed this as best they could, and made it to a farmhouse.

At North Haven the bridge was out, and there was no alternative but to ford the stream. Maxim and Law removed their shoes, rolled up their trousers, and waded in to lay stones to support their vehicle. When they finally made it across, it was half-past one, and there was no hope of reaching Branford by two. But since they still might get there in time to see the last of the race, they pressed on. They had not eaten since early morning, so they stopped and bought sandwiches to consume enroute.

Finally they arrived at Branford, and came upon a little race track, with its diminutive grandstand. This was filled with people, including many ladies, and everyone seemed very excited. They were waving handkerchiefs and programs, and Maxim looked hard to find the cause of their elation. He saw a little Stanley Steamer at the starting line, and a long, low contraption off to one side of the track, with a man under it. This proved to be a single-cylindered Winton. A man came running up to the Mark VIII and wanted to know if he was addressing Hiram Percy Maxim, of Hartford. Assured that he was, the man explained that the crowd had been waiting for him all afternoon, so that they could have a race. Only these three cars had made it to the track, and the Winton wouldn't start.

Maxim tried to explain that he had just got in, that he had been seriously delayed and his engine needed fixing, but the man replied that the crowd had waited long enough, and that the Mark VIII must proceed to the starting line without delay.

It was a critical situation. Maxim knew that his engine needed attention, and he could also picture his boss and brother-in-law, Lieutenant Eames, with his edict that the Pope Motor Company must win. But under the pressure of the waiting crowd, there seemed to be no alternative. Maxim instructed Law to unload everything they could spare to reduce weight, while he himself ran about the little machine with an oil can, squirting frantically. Amid wild applause he got the little car going, and ran it up to the starting line.

An official explained that the race would be ten times around the track, for a distance of five miles. By this time, the Stanley Steamer had so much pressure that it seemed on the verge of blowing up. It was apparent that its driver hoped to lunge so far ahead at the outset that the Mark VIII would never be able to close the gap.

The Stanley Steamer assumed the pole as the starter's

pistol shot cracked above the dreadful racket. Maxim manip-
ulated his gears as rapidly as he dared into high, but the Steamer's driver had only to open his throttle to zoom ahead. Maxim was barely to the three-quarter mark when the Steamer
shot past the grandstand, urged on by the cheers. When Maxim came along, they greeted him with jeers, and crude suggestions about getting a horse. On the next round Maxim gained
a little, indicating that the Steamer's pressure was not holding
up, and that the Mark VIII was gaining on it. The next time
around it was Maxim who received the cheers. It was now a
question of whether or not the Mark VIII could hold together
for the remainder of the course. Maxim drove carefully,
avoiding undue strain on the curves, and hugging the pole.
Every time around he gained on the Steamer, and the crowd
roared its approval. It was a real race. On the third mile,
Maxim caught his rival, right in front of the grandstand, which
promptly went wild. At the end of the backstretch, Maxim
closed in to pass, and took the lead. At the finish, the Mark
VIII was an eighth of a mile ahead. The crowd cheered its
tribute, and someone took a photograph, showing the flagman,
and the grandstand.

Maxim turned around and drove back to the line to greet
his defeated opponent, who shook hands, and offered his sincere congratulations. Such was America's first motor car
track race, with Maxim the winner.

Thirty years later there was an important conference
held in Hartford on the subject of molded bakelite products.
HPM attended it, and in the course of the sessions, was approached by a man who asked if he recalled the history-making
race. Maxim admitted that his memory of the event was still
vivid. With a laugh the man concurred, explaining that he had
been the driver of the Stanley Steamer.

In that year of 1899 the Pope Manufacturing Company,
under engineer Maxim, built the Mark V Columbia Electric
Victoria. It could travel at twelve miles per hour, and was
conceded to be the most beautiful horseless carriage ever
built. The driver sat on a high rear rumble seat, and looked
out over the top. Despite its success, faster gas engines

ruled out its production.

By this time motor vehicles began to take on aspects of our present design and arrangement. The gas pedal had a steering wheel, float-feed carburetor, spark control, brake pedal, gear shift lever, clutch pedal and foot throttle. Production officials finally recognized that even in design, the carriage had become a machine.

Finally, in 1900, the gas engine had reached the point with several builders where it gave promise of eventual success, and consequent adoption by the general public. It was then that Hayden Eames and Mr. Day sent out a frantic call for Mr. Cuntz, their patent attorney. They told him that they wanted to know what the Selden Patent was all about!

Since this was the situation he had been vainly trying to point out to them for the past two years, he became quite eloquent. He knew every word and every punctuation mark of that patent, and went over it thoroughly. The result was that Day and Eames decided to investigate the matter further. To that end, they sent Maxim to Rochester to see Selden. That gentleman was a clever patent lawyer. He had applied for his patent in 1877, but since he knew that according to the laws of his day it would run out in seventeen years, at the too-early date of 1894, he decided to keep the patent alive by asking for one impossible thing after another. He had kept this up until 1895. Then, seeing that the day of the gas engine had finally come, he permitted his application to issue. This gave him a monopoly that would be in effect until 1912. It was clever maneuvering, but of course would not be permissible under today's laws.

When Maxim returned and reported his findings to Eames and Day, they contacted Selden and bought out the exclusive rights to the patent. They then issued sub-licenses, for a fee, to the other leading builders of gas engines, and together they formed the Association of Licensed Automobile Manufacturers. One of the gas vehicle producers who showed a great spurt of growth at this time, refused to join in with the others. Of course this left the Pope Company, which owned the patent, no alternative but to bring suit for infringement. This proved to be a long-drawn-out litigation. Among others, Maxim was called upon to testify.

Maxim was primarily occupied during February of 1900 with decisions on bus designs. He kept careful records of all of his designs and tests in leatherbound notebooks which are now in the Maxim Collection of the Connecticut State Library. These accounts show that in the electric buses, the battery, complete in its tray, must not exceed six thousand pounds in weight. The tray alone weighed five hundred pounds! To be successful the bus must average seven miles per hour, allowing for it to stop ten times per mile for a trip of 15.5 miles. At this rate, its full speed must be 11 miles per hour. This seems verging on the humorous today, but it was the beginning.

The year 1900 marked the production of another interesting vehicle, the Mark XIX. This was a real "surrey with the fringe on top." It could maintain a speed of 12 miles per hour while carrying several passengers. Charles J. Glidden, sporting goggles and a linen duster, drove one of them proudly up New York's Fifth Avenue.

By winter of this year, the Mark VII's, lot number 5, were in production. Maxim took this car on a 500 mile endurance run under extreme conditions. The route was between Hartford and Springfield, and the temperature was well below zero. He made ten of these trips, wearing a Derby hat and a coonskin coat. This was the year that Maxim received the Gold Medal Award of the Paris Universal Exposition for his work on motor vehicles.

It was in 1900 that Maxim's son, Hiram Hamilton Maxim, was born, at Park Terrace in Hartford. A family picture dated October shows the proud father holding the baby.

Later that year HPM left Hartford, and went to Pittsburg as vehicle motor engineer at the Westinghouse Electric Manufacturing Company.

*   *   *

This "turn of the century" year introduced an era of great scientific progress for radio. During the next four years, over a thousand articles on wireless were to appear in American magazines, all the way from Cosmopolitan to the Woman's Companion. The impending progress in transportation and communications was incomprehensible at the time. Civilization had never known such revolutionary changes.

It was also during this year that Hudson Maxim sold the United States Government his formula for Maximite . . . "the first explosive that could be fired through armor plate and exploded after passing through the plate."

In May of 1901, the Maxims obtained a house on Howe Street, in Pittsburg. The next two years involved a great deal of car testing, and there were many family outings and picnics.

During the fall of 1901, Maxim was assigned to drive one of the Pope Company's gasoline vehicles to the Pan-American Exposition in Buffalo. This was an endurance run, and Josephine accompanied him, as she often did. It was usual to encounter severe rainstorms and cold weather, and the early motor vehicles offered no protection against either. But the young couple, with their adventurous zest for life, protected themselves by piling on quantities of heavy clothing, and then covering themselves with a poncho. Keeping their feet warm was an impossibility, however, and Maxim recounted that sometimes he couldn't bear the pain of touching his heels to the ground after one of these endurance rides.

On September 6, President McKinley was shot while attending a concert in the Buffalo Music Hall at the Exposition. He died September 14, the day that the Maxims reached Rochester. Because of this national tragedy, their run was cancelled.

*   *   *

In 1901, from December 6th to 12th, radio achieved international triumph. A station in St. John's, Newfoundland, exchanged signals with Cornwall, England, conquering two thousand miles of space without wires. The press all over the world was wild in jubilation, and in disbelief. Radio was the main topic of conversation everywhere.

*   *   *

From 1895 until 1901, the Pope Manufacturing Company, later called the Electric Vehicle Company although it still produced gas models, led the United States in the building of motor carriages. They advanced from the primitive experiments of the Lynn trip to a full line of gas and electric motor vehicles. But eventually, in the fierce competition of private enterprise, Hartford lost out to Detroit.

By 1902 the Selden Patent case was in the courts again. The highest priced legal minds had been obtained, including an attorney from England, Mr. Dugald Clerk.

The attorney for the Pope Company insisted that in order to present his case effectively, they would have to show that a vehicle built according to the provisions of the Selden Patent would actually perform. This was a terrible mandate for the Pope factory, since the patent design was crude to the point of the ridiculous. A clever engineer by the name of Henry Cave was assigned the job. After spending a great deal of time and thousands of dollars, he finally came up with a working model. The numerals "1877," indicating the year of the patent application, were painted on its side, and it was dramatically brought into court for exhibit, actually running under its own power.

The result of this spectacular presentation was victory for the Pope Company in the infringement case, and the Selden Patent was sustained. But nobody gave a second thought to the "cocky little man" who had suffered defeat. He moved immediately for an appeal, hired the finest lawyers available, and carried the battle to a higher court.

This second round dragged on for months, until finally the Court of Appeals handed down their decision, in favor of the defendant, reversing the decision, and "breaking" the Selden Patent. As Maxim explained it, the "cocky little man" was Henry Ford.

This decision was in line with Maxim's thinking all along, as he had never felt that Selden had gone far enough with his idea to have any real rights to the completed vehicles. Selden himself had done nothing toward their development.

*   *   *

In 1903 HPM returned to Hartford, and resumed his old post with the Electric Vehicle Company. For the next several years Maxim continued his car testing and experimenting while employed as their chief engineer. A typical notebook comment of this period, dated June 17, 1903, concerning the Mark XXXVIII Runabout reads: "brakes don't hold, brake levers bend. Motor hard to oil, bonnet ugly." This analysis continues through twenty-three negative points. Then he proceeds to attack each one of these problems, individually and constructively. Another entry concerns the firm's delivery wagons, which were capable of hauling loads of from four to nine thousand pounds, up to a speed of eight miles per hour. In line with these experiments, HPM was often sent out of town on assignments. One such trip was to New Orleans to start a "faulty" car. All that the job required was the switching of two wires.

During this period, some of the motor car experience was in the line of personal pleasure. The Maxims enjoyed many machine trips to scientific points, and picnics with their friends.

An important scientific event of 1903 was the International Electrical Congress, held in St. Louis, Missouri. Papers on radio were presented, which served to inspire many amateurs.

The year 1904 continued pleasantly for the Maxims. The family album shows many pictures of them doing things together. There was a June 5 trip to Clark's Corner on a nice sunny day; an afternoon down the Connecticut in the rain; a Higginum picnic, with a little fire in the rocks; an Amherst trip, with everyone very dressed up, including their friends the Todds and the Southers; and a Watertown trip, and a Branford trip. There is a picture from Glocester, with HPM wading in the creek and picking up a laughing Josephine. More formal family scenes include Josephine dressed in a suit and a large hat, all ready for bridge whist, and an interior view of the Maxim home. This shows a beautifully appointed room of the period, complete with plate rail, floral wallpaper, and built-in China cabinets. The family is seated at the breakfast table.

While Maxim was still chief engineer of the Electric Vehicle Company through 1905 and into 1906, that firm obtained parts from the Belden Auto Transportation Company in Pittsburg, and from the Olds Motor Works in Lansing, Michigan. HPM's notes for this period are often typed, an innovation for that day. He states some of the specifications for the Columbia Electric Buggy: "Lamps 22 v.... 6 candle power batteries... bell operated by a push button in control handle... 24 cells in 6 trays of 4 each. Miles per battery change, 73.3 in third speed, 50 miles at top speed." This "top speed" was 20 mph. The total vehicle weight was 1500 pounds, of which 650 was batteries. Specifications for their gasoline machines show a gas tank capacity up to 55 gallons.

Statistics on the Mark III wagons of the period show an operating cost of $31.50 per wagon, per month. One wagon did the work of three horses, which cost 82¢ per horse, per day. This was over twice the expense of the mechanical vehicle.

*   *   *

Radio's 1905 milestone was the invention by Lee DeForest of the 3 element vacuum tube. In 1906, through Richard and Dunwoody, came the crystal detector.

*   *   *

Meanwhile, the Maxim family had moved to 550 Prospect Avenue, in Hartford, and a daughter, Percy Maxim, was born there July 4, 1906. For a number of years HPM enjoyed telling his little girl that the Independence Day fireworks were in celebration of her birthday. Maxim kept her baby book personally, and recorded the list of gifts, and names of the nurses. He also included a photograph of the doctor, the congratulatory telegrams, and Percy's horoscope, a gift from her Aunt Florence.

The year of 1906 was Maxim's final one as a vehicle engineer. The motor car probably had no greater champion than HPM, yet not even he was able to see the extent to which it would become popular. He often commented that one of his deepest regrets was that Colonel Albert A. Pope, the "father of good roads" in the United States, could not have lived to see his dream materialize. Prophetically he had told Maxim one day, "I believe this horseless carriage business will be one of the big businesses of the future."

Hiram Percy Maxim had a great deal to do with the inventing and engineering which eventually produced our modern motor car. He developed the multiple-splined shaft, and had the steering wheel located on the left side instead of on the right. He solved many problems involving bearings, wheels, tires, mufflers, batteries, battery-charging, carburetors, spark plugs and brakes. Forty-nine motor vehicle patents were issued in his name.

In addition to his practicality, Maxim had an unusual sense of humor, inherited from his father, but more kindly. It is fitting that one of the final records of HPM's motor vehicle activity should be in this vein. He composed the program for the February 20, 1907 Mark III Moose Dinner at Parson's Theatre Cafe. It reads:

MENU

Anti-freezing solution

Blue punctures on the half shell
Salted split pins... pickled castings...bleached motor brushes

Venison stew

Moose steak

Double-acting jaw type, leather faced
Brazed spuds

High tension brew

Multiple jet feed   splash lubrication

Venison roast

carbonized and case hardened   alternating current jelly

Mark LXVI combination salad

Heavy Valvoline demi-tasse

cheese binders, guaranteed 60 days

Hunyadi Janas repair kit

In the early part of 1906, Maxim had gone into business with Thaddeus Goodridge, his best man, with the objective of producing and marketing an electric automobile. HPM did all of the designing at his Prospect Avenue home in Hartford. But during this research, his attention was drawn to firearms, possibly through his father's invention of the machine gun. Working with gasoline machines had involved him in developing a muffler to control the noise. These two areas began to merge in his mind, and the idea of a silencer was born. The specific solution came to him one day as he was watching the water going down the bathtub drain. The basic principle was "of catching the gases as they leave the muzzle of a gun in chambers disposed around the bore, in which the gases of the discharge would be trapped and made to whirl violently, being unable to issue into the atmosphere until their pressure and velocity had been largely dissiapted, under which circumstances little or no report or noise was made."

Maxim put in long hours at the designing board, in his home, but there was time for family activity, and for friendships and fun.

On June 3, 1907, in St. John's Episcopal Church, Percy Maxim was baptized. The Maxim album contains a sequence of baby pictures, featuring Percy in a wicker carriage with a ruffled parasol top. She is attended by Mrs. Maxim, who is wearing glasses for the first time.

A society item of the day concerned Florence Maxim, a graduate of the New England Conservatory of Music. She had just written an operetta, which was being produced in Hyde Park.

*　　　　*　　　　*

Early in the twentieth century, a young man, an amateur radio operator named Roland Bourne, had moved to Hartford to work for the Tuska Company. He met HPM, and the resultant friendship developed into a lifelong association. Maxim hired Bourne to work for him, and they combined experimental and inventive business with a great deal of fun.

The two men, Maxim and Bourne, had a foreman named Flannagan. Probably confused by the experimental nature of company business, he quit frequently. The first indication of this would come as he took off his overalls. But Maxim was rated as a good boss. He was never critical, and always had confidence, even when it came to borrowing money to back up his ideas. He was thorough and systematic, often stating that "the most important thing on a piece of paper is the date."

Maxim and Bourne always took their humor where they found it. One source of their amusement was a sign in the Porcupine Boiler Company, Williamson-Fairfield. It was a motto, done in beautiful Old English script and framed in color:

"Count that day lost when the low descending sun
Sees prices shot to hell, and business done for fun."

Maxim was always interested in the mysterious and the occult. He had a friend named Dr. Wells, a psychologist at Hartford Theological Seminary. Through him they heard of a famous medium, Marjorie, who claimed great powers, and had as her "control" a certain Walter, of Buffalo. She had been investigated by a South American committee, and by Houdini. HPM decided to visit her, and had her read his tea leaves. He discounted most of it, although she did appear to have some information. In order to convince Maxim, she suggested an experiment. He was to seal an identifiable object in a box, and within twenty-four hours she would have it out, and under the doctor's pillow.

Accordingly, HPM tore out the center of an old worthless stock certificate, and wrapped it in asbestos. He and Bourne constructed a box for it. This was a four inch cube of sheet steel, welded shut, with no door, and no lid. They covered the entire surface with a welding bead, and had a photographer shoot it from all sides. Bourne had also included two Zodiac signs, and the letter "M" for Marjorie, but they never heard from her again. Later she challenged General Electric. They enclosed a penny in an electric light bulb. She didn't solve that situation either.

As the two young men went about their designing, they enjoyed the current story about Maxim's illustrious father. He was an excellent draftsman himself, and had come into his shop's designing room one day to observe the work of an employee. The latter was doing an ink tracing on excellent, expensive linen, with a pale blue matte surface on one side, and shiny on the other. It was dusted with magnesium dust, and truly material for an expert. Hiram Stevens Maxim bent to study the drawings and spat his wad of tobacco juice right into the middle of it. That was his way of delivering an opinion. It was never HPM's way. He raised questions, or made tactful suggestions, which habit was to earn him the designation "the beloved gentleman."

*   *   *

During this time, amateur radio was improved with the use of tuners, and Bourne got his first two-way rig. His activities were beginning to influence HPM in communication experiments.

After Bourne had been in Maxim's employment for about a year and a half, and being very conscientious, he went to his boss and tried to resign. He said that he had been trying to "get the feel of the job" but didn't believe that he was contributing anything. Maxim proved that he was a very understanding man. "Keep on doing what you're doing," he replied, "and we'll talk about it six months from now." Bourne was to continue in Maxim's employment for over thirty years!

Maxim made the drawings, applied for the patent, and manufactured his first gun silencer in 1908. It still needed work, and was perfected in 1909. The first Silencer was tried on a .30-.30 Winchester. With the Silencer, .22's could be made almost perfectly noiseless.

*   *   *

On January 2, 1909 the first American radio organiza-

tion was formed. It was called the Junior Wireless Club Ltd.,
of New York, and was composed of five boys with an average
age of twelve years, with their parents.

Later that month the Wireless Association of America
was organized, with a membership of 3200 amateurs.

Several months later, in 1909, HPM built the Maxim
Glider. It was flown as a kite, with a line from the outboard
strut of each wing, to sticks in the ground. There were no con-
trols on the machine, and balance was maintained by the shift-
ing of the pilot's weight. It was a Chanute type, made of knock-
down parts purchased from Witteman Brothers of Staten Island.
They were later to develop the Barling tri-plane bomber. The
Maxim album contains a picture of HPM's glider, taken in the
fall, at North Meadows near Hartford. It was piloted by George
Lucas. Mr. and Mrs. Maxim are shown watching the flight.
He is on crutches, the result of a previous take-off in this
glider, when it was being towed by a 4 cylinder auto from the
Electric Vehicle Company.

The family section for 1909 also features a photograph
of Percy Maxim, wearing a lacy ribboned bonnet, a white coat
with large lace-edged Bertha collar, white hose, and high black
laced shoes. Following this is a picture of HPM playing baseball, a sport he enjoyed. The teams are not wearing uni-
forms. Every player is attired in a shirt and tie! This historical family album also records the acquisition and death
dates of a succession of Maxim cats, including one called Ann-
abel Lee.

These early 1900's marked a new era, Noise. It was in
everything, automobiles, gas engines, street cars, trains,
munitions, and even in the popular songs of the day. One of
the current numbers was ·"The Automobile Song," and the
"Menagerie Song." This was the age of big bands, including
Sousa's. There were "bones" and tambourines, and minstrel
shows, as America Cake-walked and Turkey-trotted into the
twentieth century. As progress increased, so did the noise,
until it became a most undesirable by-product of civilization.
It was not uncommon for the engines of some mills to torment
the ears for distances of over ten miles.

Maxim, already successful in the development of the gun silencer, began to consider the application of its principles in other fields. He believed that noise could be enormously detrimental to the nervous system, and regarded the control of noise as one of the major problems of the day. He wrote numerous articles to this effect, and they were published in various national magazines.

His first application of the silencer idea came with the invention of the car muffler. This he proceeded to adapt to stationary gas engines, motor boats, steam exhuasts, air conditioning and ventilating systems, Diesels, and many other phases of industry. He once made a trip to Tucson, Arizona to quiet a factory that could be heard for fourteen miles. He installed a silencer with a muffler diameter of five feet, at a cost of six thousand dollars, a huge sum for those days.

Maxim said of his company, "We believe we have spent more money and more effort in the study of NOISE REDUCTION than anyone else in the world.

There had been much misunderstanding of the Maxim Silencer on the part of the general public. Some people immediately envisioned a host of crimes, perpetrated and made possible by the use of a gun with a silencer attached. The idea became the basis of plots for countless mystery stories, and was referred to as a menace to public safety.

Nothing could have been further from Maxim's intent. The company's information pamphlet stated: "Maxim Silencers cannot be used on revolvers, shotguns, automatic pistols, nor automatic rifles of larger caliber than .22."

The primary purpose of the gun silencer was its use by the United States Army. Official tests were conducted with the apparatus at Fort Snelling, with the resulting report that it offered many advantages. First of all, it facilitated the instruction of recruits while firing, since the voice of the commanding officer could be heard clearly at all times. This was important in a time when there was no electronic field communication. Moreover, the Silencer permitted a lessening of strain on the nervous system. Chronic flinchers became good marksmen. Recoil was lessened, and therefore fatigue was reduced.

62

There were tactical military advantages to the Silencer too. The lack of noise made it possible for men to conceal their positions, and the location of their firing line. Since the use of the Silencer annulled the flash, it also helped prevent disclosure of position during night firing. It is understandable why the use of the Maxim Silencer for machine guns was officially recommended for the United States Army.

But always, Hiram Percy Maxim was a man of peace. In fact, his greatest hope for amateur radio, which he was later to organize and help preserve, was that it might be the means of uniting all the countries of the world in understanding, and consequent lasting peace.

By the year 1910, the importance of radio was being recognized. On June 24th Congress passed a bill requiring communication apparatus and operators on our ocean liners.

Up until this time, practically all radio had been amateur... from amator, lover... for the love of the work and the thrill of achievement. The word Electricity had a classical derivation, too, coming, in 600 B.C. from the Greek word for amber, a substance which attracts. The term appeared scientifically in a book by Sir Richarde Browne in 1646.

In 1910 even Hudson Maxim, the inventor of smokeless powder, had turned to gentler activities. He wrote a constructive book titled The Science of Poetry and the Philosophy of Language.

In 1911, inspired by the activities of Roland Bourne, Maxim became an amateur radio operator, along with his son Hamilton, who was eleven. His first station was SNY, then 1WH and 1 ZM. It had a range of one city block. After World War I, to his final QRT, Maxim was 1AW.

Some of HPM's experiments in this period had their humorous side. He kept a barrel of castor oil in his garage, for lubricating machine parts. One day he sent little six year old Percy out to bring him a cup of the nauseous liquid. Fearing that her father was planning an unpleasant medical experience for her, she pulled the plug on the barrel, and let the contents spill all over the garage floor.

In February of 1912, Josephine Hamilton Maxim received

a notice of membership in The Connecticut Society of the Co-
lonial Dames of America, through Tristram Coffin, the New
England author. This marked the beginning of her participa-
tion in the affairs of her country, which interest was to involve
her in politics and progress for many years.

By this time, Maxim had expanded his Sil ncer business
into many new fields, including steam engines, and boat engines. A "Silencer" for a Model T Ford cost six dollars.

*   *   *

HPM had improved his amateur station so that it now
carried for five miles. Larger ones were becoming capable
of transmitting for over a hundred miles. But this distance
was still far short of the dreams of the "hams," a term derived from the word **amateur**. All across America, men and
boys with an electrical bent were working with wires and coils
and miscellaneous junk in their own private quarters. Many
leading inventors and engineers of radio got their start in this
manner, as amateurs. Prominent among them were Pickard
and Fessenden in the United States, and Poulsen in Denmark.

These early pioneers of wireless had many difficulties.
There was a lack of information and correlation. It was hard
to obtain supplies, and there was rivalry among the various
firms. There was no organization, and operators used any
wave length, any power, and any type of transmitter they
wished, with the result that before long there were serious
conflicts with the Army, the Navy, and the Department of Commerce, which had also begun to get into the act.

The battle of the wave length raged in Congress. Governmental and private interests fought for choice spots on the
band, culminating finally with the passing, on August 9th, of
the Radio Act of 1912, signed by President Taft. By it, amateur radio was not abolished, but it was restricted to 200 meters, which was an area deemed worthless at the time. Later,
this was to prove a fortunate protection, as it gave amateur
radio a legal status. Maxim called the Act "one of the most
constructive and valuable bits of legislation that a Congress

has ever enacted."

This law that was designed to condemn the amateur to oblivion, instead opened a challenge. Five thousand clever and ambitious young people were not to be easily disposed of, and they set to work to find a way to utilize their assigned territory. They were spectacularly successful, and the handling of messages for fun, for friends, for help in emergencies, became their dominant activity. This was a historic development, for it marked the birth of short wave radio. The resourceful amateurs, working for the pure joy of it, took their super-high frequencies and developed them into the longest distances we have on earth, or indeed, in space. In March of 1913, amateur radio received public acclaim for help in an emergency, a devastating windstorm involving Ohio and Michigan. Such humanitarian dedication is the ham's greatest satisfaction.

January of 1914 was a month of destiny for amateur radio. It was then that the first meeting of the Radio Club of Hartford was held. Presiding on the historic occasion was Hiram Percy Maxim. A "blue-eyed boy of eighteen," Clarence Tuska, was appointed as secretary.

HPM, the brilliant automotive engineer, was at it again, in a new field. He had a one-kilowatt station, with a maximum range of one hundred miles, if conditions were favorable. One April night, after he had been trying unsuccessfully to locate some vacuum tubes he needed, he was struck with an idea. He would try relaying a message concerning the parts he required. The plan succeeded, and he thought deeply about the matter the next morning as he drove his big automobile to his downtown office.

Not that the idea of relaying a radio message was new... ships at sea had been doing it for a long time. What Maxim had in mind was something bigger than that, a national organization for relays. Organization was the key word that would join all the amateurs in the country into one strong, capable body. He even thought of a name for it... the American Radio Relay League.

Since Maxim was not a man to let a good idea lie for

very long, he presented his suggestions to the Hartford Club
at its next meeting. The members, together with the children
in their families, spent many evenings writing letters, and
stuffing envelopes. By the middle of May, hundreds of applications had been mailed out. There was to be no membership
fee, but the qualifications were high, so that only the best amateurs would be included.

The response was a landslide of memberships, and by
June, relays were being organized from between such distant
points as Boston and Denver. It was Maxim's dream to improve all this until it became a genuine service, both personal
and public. By August of 1914 there were more than two hundred designated stations. By September they covered thirty-
two states and Canada. In October the club issued its first
Call Book, listing all the amateur stations. By this time the
operation was so large that it needed a fulltime head, and the
organization hired Ken Warner. Late in 1914, Maxim went to
Washington D. C. , to confer with the Commissioner of Navigation of the Department of Commerce. His purpose was to gain
recognition for the club in official circles, and to get special
licenses for certain relaying stations.

Maxim, a born organizer, had other creative activities
too. He, together with Tuska, Bourne, and Charles Barrett
were promoters of the Hartford Writers Club.

It was about this time that the newspapers of the world
featured a photograph of Maxim's father, wearing the Cross of
the Legion of Honor, and numerous other decorations. He was
extremely distinguished-looking, with thick white hair, a neat
moustache and beard, and a proud, direct gaze. He had been
knighted by Queen Victoria for his invention of the machine
gun, by means of which England had won the Boer War.

In February of 1915, Maxim left the Hartford Club, and
incorporated the American Radio Relay League. This organization grew with pleasing rapidity, but was suffering from a
lack of funds. With no dues from its members, and pamphlets
and bulletins being issued at cost, there was no way of financ-
ing the organization. The only solution which President Maxim
and Secretary Tuska could come up with was the publication of

a magazine. Acting immediately upon the idea, even to the point of digging into his own pockets to help finance it, Maxim became the guiding genius behind the amateur radio magazine, QST. This publication, sold by subscription, made its first appearance in December of 1915. There were 24 pages, with a blue paper cover designed by Tuska's uncle. The price was 25¢ for a 3 month's trial subscription. The first business manager was Tuska, then a student in college. The number of pages in each issue depended on the amount of cash receipts on hand.

By this time it was "amateur radio," not "amateur" radio. The difference was that the random experimenter who had been fooling around, without a destiny, had pooled his information and become organized. Yet at this point even Maxim would not have cared to predict out loud that some day the entire world would be hooked up into an international traffic route.

To improve transmissions, Maxim planned six trunk lines to cover the United States, and the first country-wide relay was held on Washington's birthday in 1916. This operational occasion involved the governors of all the states, and President Wilson. The message for the occasion, perhaps reflecting affairs in Europe, was: "Democracy requires that a people who govern and educate themselves should be so armed and disciplined that they can protect themselves."

By 1916 the Maxim family had moved to a pale yellow two-and-one-half story house at 276 North Whitney Street, in Hartford. There was a greenhouse and a large garden in the rear, and eventually, a Packard in the driveway. Across the street was Elizabeth Park, where Maxim tried to teach his little daughter how to play baseball like a boy.

In a projecting windowed area of the house, referred to by the family as "the conversary," Maxim's rig stood on a table against the wall. This was Old Betsy. She ran 7000 rpm, belt-driven by a 1/2 horsepower motor in a box in the corner. It featured a rotary spark gap, and two large oil drip cups, which needed frequent refilling. HPM was on the air with this rig from 7:00 P.M. until 1:00 A.M. every night. This was the period when Maxim was regularly using his Franklin car to truck 5 or 6 mailbags of QST's to the Post Office. It also marked the year of the first amateur radio communication with a plane in flight.

While they were living on Whitney Street, the Maxims took time out for a vacation trip to Cuba. They went by train to Key West, and from there they took the boat for Havana.

In England, in 1916, Sir Hiram Stevens Maxim died. He had lived seventy-six brilliant years. Only a few months previously he had been shown in a movie newsreel reciting "Mary Had a Little Lamb" in an impressive chamber suggestive of the British House of Commons. His purpose was to prove that it was not what you say, but how you say it, that is important. Sir Hiram had become a rich and famous man, but as his son, HPM put it, "He never quite learned how to be a father."

Uncle Hudson, in 1916, was still going strong. He had just published a book called **Dynamite Stories**, and appeared in his 'longies' as King Neptune in that year's Atlantic City Beauty Pageant.

The only joint venture these two genius brothers ever accomplished was the invention of a combination coffee pot and tea pot. This strange, ungainly contraption was designed as a small container inside a big one, so that the heat from the boiling coffee warmed water for the tea to steep. It achieved neither recognition nor profit.

*   *   *

HPM, as the founding father of the American Radio Relay League and of its publication, QST, wrote its chief editorials. But the big secret of the magazine was that HPM also wrote its popular feature, T. O. M. Only six people ever knew that he was the author, until after his death over thirty years later. The initials stood for The Old Man. In contrast to HPM's scholarly editorials, these articles were earthy. Their purpose was to reach the young operators and get them to correct their bad habits. Maxim was enough of a psychologist to know that this could never be accomplished by preaching, so he wrote humorously. Hams turned to the T.O.M. column before they looked at the rest of the magazine. They read, they laughed, and they remembered. A sample of The Old Man's style may be found in the January, 1917 issue: "Rotten construction, Rotten ground leads, Rotten QRM." He goes on to complain about some Fort Wayne signals that had been cluttering the band...

"Whose undershirt they were talking about, and what schlipshung one over is I do not know, exactly, although I have a rough idea... his rotary has a bad case of the ugerumph, and somebody around the place must have spit on his rettysnitch... Rotten tuning... I wish I was a radio inspector... I would make him eat his set before my eyes and not finish until he had gnawed off the pole at the butt end and swallowed the chips. Rotten fists... Rotten!"

As the ranks of amateurs grew rapidly, manufacturing firms sprang up to supply them. These ambitious communications experimenters now had a new objective... a transcontinental relay. This they accomplished on January 17, 1917, with HPM being one of the leading participants.

On February 6th the amateurs scored an even greater accomplishment. They started a message on the East Coast, relayed it to the West Coast and got an answer back, all within an hour and twenty minutes. This historic month of February also saw a change in the organization of the American Radio Relay League, expanding its governing structure, adopting a constitution, and outlining policies.

Then in April of 1917 there came a ruinous blow, in the form of a letter from the Chief Radio Inspector of the Department of Commerce. A copy of it was mailed to all licensed amateurs in the United States. It said in part:

"I hereby direct the immediate closing of all stations for radio communications both transmitting and receiving, owned and operated by you. In

order to fully carry this order into effect, I direct
that all the antenna and all aerial wires be immediately lowered to the ground and that all radio apparatus... be rendered inoperative... Immediate
compliance with this order is insisted upon and will
be strictly enforced. Please report on the enclosed
blank your compliance with this order; a failure to
return such blanks promptly will lead to rigid investigation."

It was a time when strong men cried.
Immediately following this catastrophic order, a bill was
introduced in the House of Representatives proposing that all
radio communications, including amateur, be turned over to
the Navy. Again, Hiram Percy Maxim went to Washington.
The bill was killed in committee.
Then the United States entered World War I. Maxim's
Silencer Company turned to the production of munitions, including bayonet scabbards, and gas grenades.
The first big problem the military leaders encountered
was a complete lack of radio men, from instructors to operators. The situation was urgent and critical. Washington contacted New York. An official there wired Hiram Percy Maxim.
The call was for five hundred trained operators at once.
Also, could the radio equipment of the amateur stations be
converted to military use? Ten days were allowed for this to
be accomplished, and within that time, the Navy had its quota
of experienced operators. The next call was for two thousand
radio men. They too were enlisted almost immediately. Over
four thousand of them saw service before the war was over.
But still the ban on amateur radio was continued.
In order to maintain home training, the amateurs organized a radio class for ladies. HPM's wife enrolled, and they
met daily at five o'clock in the Maxim home. Their instructor
was Miss Cecil Powell, the secretary of the Maxim Silencer
firm. There were sixteen members in the class, ranging all
the way from debutantes to matrons. They sat in a little circle, much like a kindergarten. Miss Powell had a buzzer on a

board, with a telegraph key and two dry cells, for code practice. A newspaper picture of the group shows them attired in
light-colored summer dresses, with sweaters or coats, and
big wide hats. The caption reads, "Mrs. Maxim is the wife of
the well-known inventor and munitions maker, Hiram Percy
Maxim. This is the first wireless organization to be founded
by women in this country." Maxim himself commented that
they had learned as much in seven lessons as a class of men
in four months.

The Maxim children, Percy not yet a teen-ager, helped
out in the war effort by working in their father's factory. Percy's job was packing hand grenades. The following year, young
Hamilton Maxim, a student at M.I.T., joined the Navy Training Corps.

The Armistice came November 11, 1918, and the radio
amateur had played a large part in helping to achieve it. But
still the ban prevailed. In Washington, an attempt was being
made to give the Secretary of the Navy control of all radio in
the United States. The Board of the American Radio Relay
League, with a capital of ninety dollars, which the members
themselves put into the pot, delegated Maxim, at his own expense, to attend the hearing on the bill.

In addition, the League issued its famous "Blue Card"
appeal to every amateur who had been licensed before the war
began, begging for their support and assistance. Maxim was
the author of the cards. Thousands of answers poured in,
some from members of families whose relatives had been
killed overseas in line with their communications duty. The
Cards were a great gesture, and an example of inspired lead-
ership. The threatening bill died in committee. To help handle the increased work of the League, the group hired Ken B.
Warner as their executive secretary. He was fresh out of the
Army, in fact, still in uniform.

In March of 1919, the League met, and with only thirty-
three dollars in the treasury, voted to resume publication of
QST. The eleven Board members present, including Maxim,
put up their own funds to finance the cost of the new issue.

Also in March, on the 24th, Maxim's mother-in-law,

Mrs. Hamilton, died, after five days at Mercy Hospital in Baltimore, Maryland. She had contracted pneumonia after a scalding accident at the Rennert Hotel, in Baltimore. Eighty-two at the time of her death, she had, as noted in the papers, once "graced the Governor's mansion as hostess... suave and elegant."

*   *   *

On April 12, 1919, the Navy Department announced that effective on that day, the ban on amateur receiving would be lifted, but that restrictions against transmitting would remain in effect until peace had officially been declared by the President.

Clinton B. DeSoto, the recognized historian of amateur radio, describes what ensued:

> "Thousands of amateurs throughout the nation rushed frantically up to long-deserted attics or down to musty basements where the old apparatus lay, intact under its seals, in cobwebby, dust-covered decay. Hastily it was brushed off; tenderly idolatrous fingers carried the individual units to old resting places; tremblingly, bell wire was stripped of its insulation and connections wired in place. The towering antenna of old, dismantled in 1917, was mourned for a bit, in silence; and then work started on a new network of wiring, to be strung gingerly aloft from tree or roof or mast. Hungering, code-sick ears, sad in the nostalgia of two long weary silent years, absorbed in ecstatic reunion the roaring threnody of the commercial and government stations."

But there was yet much work to be done. The League issued bonds for more funds, and the first post-war issue of QST was published. But amateur radio was still in a sorry situation, being without authority to transmit. Appeals and

protests besieged Washington without result. Indeed, there
came a new threat. It was being urged officially that the Navy
Department be given a monopoly on all oceanic and international radio.

In August of 1919, with the Secretary of the Navy absent
in Hawaii, the Assistant Secretary, Franklin Delano Roosevelt,
stated that he did not know why the ban had not yet been completely lifted... that it was still awaiting action from Daniels,
and a proclamation by President Wilson. Maxim covers this
situation in an article in QST dated September, 1919:

> "Mr. Roosevelt stated we could be released as
> soon as Mr. Daniels would permit it, and in re-
> sponse to an inquiry addressed by a Senator, wrote
> that 'the Department has decided to remove the
> war-time restriction on radio co-incident with the
> proclamation of peace by the President.'"

> QST continues: "We pointed out to Mr. Roose-
> velt the bad odor overhanging the whole affair and
> the extreme desirability of a statement by the Navy
> explaining why we were held up and what we may
> expect, if the suspicion with which the amateur
> world regards the Navy Department is to be elimi-
> nated. Mr. Roosevelt promised to immediately
> wire Mr. Daniels, and we hope to know just where
> we stand."

On September 26, 1919, the Director of the Navy Communication Service announced the lifting of the restrictions,
and the resumption of authority thereof by the Bureau of Navigation of the Department of Commerce. This was complete
victory. The October issue of QST carried a one-page sup-
plement on pink paper, announcing "Ban off! The job is done
....coming...the greatest Boom in Amateur Radio History!"

The November issue of QST reported on the triumph in
detail, saying, "The incident is an example of the value of organization." This was a prime Maxim principle. The article
expressed appreciation to Congressman Greene, "Chairman of
the House Committee on the Merchant Marine and Fisheries,

loyal protector of amateur rights, co-founder of all government ownership programs, and for the past twenty-one years
Representative in Congress from the State of Massachusetts...
We found a real sympathizer in Mr. Greene. Once he had been
informed of the situation he went to work, and within a month,
Presto! We are open!... We all feel a deep sense of grati-
tude to Mr. Greene for his splendid efforts in our behalf."

Resumption of transmitting was slowed by the problem
of licensing. During the two and one-half years that the Navy
had been in control, all amateur licenses had expired. It was,
therefore, necessary to issue new ones, under the Bureau of
Navigation, which was, unfortunately, short of clerical help at
the time. However, a system of temporary authorizations
was worked out to bridge the gap. By November of 1919, the
radio amateur was back at his transmitting key. In December,
Hiram Percy Maxim participated in the first post-war transcontinental relay. He had been assigned the first call letters
of the new system, 1 AW.

It was in this year of 1919 that QSL cards were introduced. They were the idea of QST's staff cartoonist, Don
Hoffman.

A boom in the manufacturing of radio equipment developed, until it was almost impossible to supply the demand. A
new industry had been born.

But Maxim's own business was at a complete standstill
after the war, and the consequent cessation of the munitions
making. His son Hamilton had just been graduated from MIT,
and the Silencer Company consisted solely of HPM, Hamilton,
Miss Cecil Powell, secretary, and one lone stenographer.

Inspiration came to Maxim in a hotel room one night in
Milwaukee. Later he wrote, "It was so noisy when I opened
the windows, and so hot when I closed them, that I couldn't
sleep. I decided then that noise was man's worst enemy, and
started to work on this."

The result of this experience was the eventual invention
of air-conditioning apparatus which proved to be the salvation
of the Silencer business. "Noise," Maxim wrote, "causes

more damage in this life than fire does."

About gun silencers, which he no longer made, he said,
"We stopped before crime and street shootings became so frequent. Now you know why I'm glad." He went on to relate how
some hard-faced fellows once came into his Hartford office offering one hundred dollars for a silencer. These had sold for
five dollars and ten cents when they were on the market.

With his usual tenacity and organization, HPM proceeded
to rebuild the business along new and constructive lines. He
developed silencers for Diesel engines, and for factories which
were making city life almost unbearable with their uncontrolled
noise. HPM's business was soon thriving again. His office
was opposite Bushnell Park, Hartford, on the fourth floor of a
substantial brick building with a view. His son worked there,
and occasionally his daughter, whose job was to put couplings
in silencers. HPM frequently came home for lunch, and walked
a great deal. This was a writers' district. Harriet Beecher
Stowe, Henry Ward Beecher and Mark Twain all lived in the
area.

*   *   *

In January of 1920 the country was singing "The Japanese
Sandman," and the ARRL had expanded into Canada. A newspaper account of the period comments on HPM as a presiding
officer... "Keen as a parliamentarian, and judicious in allowing reasonable latitude, he has presided over some pretty interesting meetings which required one of the essential ele-
ments in a chairman—tact."

It was about this time that Maxim made a bet with his
sister-in-law, Julia Hamilton Briscoe, that he could write a
movie script and have it produced. His point was that anybody
could write a scenario. The result was a flamboyant story
with a tropical setting. No one was more surprised than Maxim when his script was accepted, and scheduled for prompt
Hollywood production. Under the title "The Virgin Paradise,"
a Fox Photoplay Production, it had its premier at the Park
Theatre in Hartford. The newspaper account referred to it as

"a story of the jungle and civilized hypocrisy... a picturesque romance which begins in the islands of the South Seas and culminates in an aristocratic old New England home."

Pearl White, one of the leading cinema actresses of the day, played the leading role. The newspaper reported her portrayal as "a waif of the South Seas.... a child living alone on a desert island through the death of her father, a missionary. Reared in these primitive surroundings, a freak of fortune makes her the heiress of millions, and she is transplanted." The article goes on to say that she rebels at her new surroundings, but eventually finds true love amid the hypocrisy.

Maxim himself referred to Pearl White as chalk-faced, red-lipped and moon-eyed, which was after the beauty standards of those days. The New York Morning Telegraph stated, "Mr. Maxim has fussed about in the past with all sorts of high explosives, but never with anything as dangerous and ignitable as a screen star."

A newspaper headline commented, "Maxim, who took the noise out of bullets, put pep into films"—giving Pearl White "an opportunity to frolic with lions and monkeys."

In the production, "Pearl engages in a rough and tumble honest-to-goodness fight with a man, and whips him! She did such a thorough job that many meek husbands were afraid to leave the protection of the theatre." The newspaper account continues, "... she was playing with some lions that were either too old to chew raw meat or too well fed to be hungry."

Pearl fell in love with a man who came in on a schooner and ashore in a skiff, and tried to lock him up in a cave. When he escaped, she swam out to the sloop after him. "Then the captain tried to marry her to a villain with well-oiled hair. The man who had vamped Pearl rescued her from that.... but even in the Long Island home that was now hers, 'the villain still pursued her'... much furniture was broken, many clothes were torn." After Pearl "delivers a little speech on the hypocrisy of civilization she allowed the place to burn down, found the man she loved, and lived happily ever after."

This wild story, concocted in a satiric spirit of pure fun to win a bet, was treated seriously in magazine and newspaper accounts... "Tonight's picture is by HPM, the handsome old rascal who invented the Maxim Silencer... he has a private wireless station at the rear of the Maxim home... two big poles with over a dozen wires strung between."

There were special souvenir programs for the premier, and Maxim and his family were honored guests.

It was also about this time that the movie, "Within The Law," was released, after HPM had given special permission, and specific instructions, since the plot hinged in large part on the use of the Maxim Silencer.

Many of the Maxims were interested in acting, some strictly as amateurs, and some as professionals. Clare Eames, the granddaughter of the late Governor Hamilton, and daughter of Hayden Eames and Clare Hamilton Eames, became a well-known actress. She appeared with Ethel Barrymore, played Mary Stuart in Drinkwater's "Mary, Queen of Scots," and appeared as Queen Elizabeth. A cub reporter who interviewed her, unused to stage makeup, "was tempted to ask her if she had tried iodine for her bruised eye." The Maxim album contains many photographs of this charming actress.

In this same busy year for the Maxims, Josephine, a born Democrat through the Hamilton family, launched her suffragist activities. As a pioneer in this field, she ran for office, and was elected to the Hartford Board of Education in May. That same year, she introduced Presidential candidate Cox when he came to Hartford.

In June of 1920, Mrs. Maxim went as an alternate delegate to the Democratic National Convention in San Francisco. This was the first time a woman had been chosen in Connecticut for such responsibility. HPM was selected to go along as a newspaper correspondent, his daily articles appearing under a by-line. HPM himself had Republican leanings, but Josephine commented that his association with the trip was "not as a silencer, but as a paying guest."

The delegation traveled by train, and the trip, plus the convention, occupied much of June and half of July. The newspaper, in commenting on Maxim's coverage, said "... his quaint wit and innocently-phrased humor are delightful and refreshing." You can... "read every line of HPM's letters and chuckle over their humanities."

Mrs. Maxim set a precedent for Connecticut women by attending this convention. Having been a leader in the suffrage movement in Connecticut, she stated, "Women should take their place in running the affairs of the country." She also was a member of the Democratic State Central Committee, and was associated with the Red Cross. She worked at their information desk. Later she became the first President of the Hartford League of Women Voters.

As the convention delegates traveled through the Grand Canyon on their way to the Coast, Maxim began his column: "With the Connecticut Democratic Delegation En Route." Using the famous quotation about the Grand Canyon, "God's boldest and most flaming signature across the earth's surface," he asserts that a visit to the Grand Canyon and a Democratic convention don't mix, because it requires too much time to see the great natural wonder, and they have only a day. He expresses a desire to return sometime and view it with his son.

In this somewhat desert area, Mayor Dunn, a Judge from Willimantic, Connecticut, barehanded caught a live toad. He put it into a cigar box, and was occupied for the rest of the trip in finding flies to feed it.

One delegate became slightly ill; one had a sprained knee; one developed the toothache. HPM commented, "Dentists are not very thick around these parts." Trying to help, he located a young woman who was once a nurse. Maxim adds, "I suggested the name of a very excellent horse doctor who lived nearby."

Near Gallup, some Mexican bandits held up the train. Two were killed, and the rest fled. HPM commented that "Since delegates pay their own way, the bandits thought the train was well-stocked with legal tender."

On June 23 the delegation was traveling across the Mohave Desert. It was one hundred degrees in the Pullman when the train was moving, and well over that when it stopped. Maxim wrote, "The desert seems a ghastly place to one from Connecticut's green hills and moist meadows.... water must be hauled hundreds of miles, otherwise there is none for locomotive, man or beast, and yet, in spite of it all, the automobile comes."

The etiquette of those days required that men wear their suits when traveling by train, but Maxim reports that on this torrid occasion, one of the men dared decorum, and took off his coat.

Mrs. Maxim, contrary to her husband's tastes, played bridge in the Maxim compartment for four hours in one hundred plus temperature and lost, paying her 5¢ debt and getting a duly-executed receipt. Meanwhile the judge was still looking for flies for his toad.

By June 24, the delegation had reached Los Angeles, and HPM commented, "It is strange how quickly one comes to know well one's traveling companions."

The delegates visited the Fox Motion Picture Studios (the place where Maxim's scenario had been produced), on a tour provided through the courtesy of the Los Angeles Chamber of Commerce. HPM commented in his column on the Hollywood actresses: "They do not paint themselves yellow, red or green, as we have been told. They whiten their faces and strengthen their lips and eyebrows. Since it is so common nowadays to see so many of our young ladies who have fallen into the same error of judgment, it does not strike one as especially unusual."

HPM continued his impression of California: "I have been dragged into and out of churches, motion picture studios, golf and country clubs, gorgeous estates, trolley trips, patios ... and indescribable eating places where they expect one to eat indescribable mixtures that taste of onions and red peppers, until it is all one confused nightmare... I'm sure I'll remember Hollywood; everything is only 7/8 of an inch thick."

HPM got his first view of the Pacific Ocean, but was more concerned editorially with the income tax. "... this is the place where most people elect to come and spend their declining years, and what may be left of their fortunes after paying income taxes."

HPM, eating on board a bleached old sailing ship at Venice, California, comments on convention luncheons as being intensely interesting and enjoyable, with their laughter, wit and repartee. He wrote for his newspaper, "These Democrats are bright and witty people." And slyly, with perhaps a Republican chuckle, he says of one of the delegates "... he remains cheerful, being a Democrat." He adds, "For the first time in history, women will be considered as good as men in a national convention." This was the group that nominated Cox to oppose Harding for the Presidency.

The personal highlight of the trip for Maxim was his visit to the San Francisco Radio Club, the first time an ARRL official had been on the West Coast.

On the return journey, on July 10, the delegates stopped in Yellowstone Park, where they encountered problems at the hotel. Maxim gave these national prominence in his column. According to him, delegates were charged double for a bath, because it was located between two rooms, and each got charged for a room with bath. He reports standing in line to get into the dining room, although the tables were empty. He wrote, "A young lady was stationed at the closed doors of the dining room, and she had to be completely satisfied as to one's antecedents first. Then she had to consult in whispers with the head waiter, after which he had to make his own personal appraisal."

Maxim's general comment on the delegation praised "the elevating effect of women in American politics," and referred to the men as "gentlemen" (in this period of Prohibition)... "whose pockets bulged with expensive-looking cigars, and prepared to treat for snakebite on a wholesale scale."

On July 18 the conventioneers returned to Connecticut, having been gone for one month and one day. HPM commented for the newspaper on their arrival, "One's impedimenta swells up during one's travels... bags which closed with only a moderate degree of remonstrance simply revolted at closing the last of the trip..." Because of too much "highly colored wearing apparel, and dangling white tapes," (this last in reference to the ladies' corset strings), the "perfectly reputable suitcases presented undignified, gorged appearances."

Mrs. Maxim left the train first, at 5:12 A.M. There

were few to see her disembark, as many of the delegates went on further, but for three hours the night before there had been twelve "highly compressed people" in her compartment. And Mayor Dunn still had his cigar box with the air holes in the top, for his toad.

While the Maxims were away on this convention trip, they had left thirteen-year-old Percy with her Aunt, Mrs. Briscoe, at the old Hamilton estate, Oak Hill, in Hagerstown, Maryland. An August note from the TOWN TOPICS comments that upon the Maxims' return, Percy accompanied her mother back to Hartford, "much to the candid relief of her two aunts, who find handling business and political affairs easier than disciplining this lively young niece, who sometimes needs a Maxim silencer herself."

* * *

The July issue of QST featured Maxim's antenna system on its cover. The cables were 50 feet in length, strung between two masts 80 feet high.

By this time, the ARRL had adopted their official diamond-shaped emblem.

In August, Hiram Percy Maxim received a formal invitation for the ceremonies "attending the notification to the Honorable Franklin Delano Roosevelt of his nomination as Democratic candidate for the vice-presidency of the United States, at Hyde Park, New York, August 9, 1920."

It was also in August that the Suffragist Movement, in which Mrs. Maxim had been so involved, achieved its goal, and American women were granted the right to vote.

In September, a reporter looking for a feature story happened to discover that there were three important hats side by side at the cleaners (Don Dolittle, of Van the Hatter fame). One belonged to Connecticut Governor Marcus Holcomb, one to Congressman Augustine Lonegran, and the other was Hiram Percy Maxim's. So there was one Republican hat, one Democratic hat, and the last one... independent headgear.

In October, Mrs. Maxim ran for office again. This time

she was seeking a seat in the State Senate. She was called "a speaker of more than ordinary power and conviction," and the Hartford paper referred to "her winsome smile, and witticisms."

This year of 1920, such a productive one for the Maxims, was the spot in American history where the Jazz Age began. This initiated a nine-year spree, gaudy with boom, nonsense, rebellion and lawlessness. There were bombings, strikes, raids and riots. There was trouble with the I.W.W. and with the Ku Klux Klan. This was the time when women began to paint their faces, bob their hair, and flaunt short skirts. Their new freedom, assisted by the can opener and smaller households, took them into the employment field. They wanted independence. They had wanted and won the vote.

With 130,000 American dead in the recent war, with strife, depression and unemployment in the offing, it seemed that the voice of woman in government could not make things any worse... indeed it was highly probable that her natural devotion, compassion and sense of justice might do the country a great deal of good.

People were singing, because there was a wealth of pleasing music, including "Avalon," and "Margie." Irving Berlin had begun his ascendancy. People were dancing "That Naughty Waltz," and going to the movies to see Charlie Chaplin, Douglas Fairbanks, Mary Pickford, and Theda Bara. Jack Dempsey had knocked out Jess Willard to become heavyweight champion of the world.

In this year of 1920, Westinghouse began the manufacture of fairly simple and inexpensive radio receivers, and a number of stations took to the air, sponsored by newspapers and department stores. This new medium was readily recognized as having tremendous advertising potential, and the American system of commercial radio broadcasting came into being.

Early in 1921, Maxim became associated with a tall slim Swede by the name of John Sundkvist. He owned and ran the American Tool Works in Hartford, and became a sub-contractor manufacturing the silencers. This relationship was maintained until Maxim's death fifteen years later. On the day of his signing in, Maxim told Sundkvist that he'd never get rich on the job, and that he wouldn't even show him the old shop tools, because "they're so bad they'll corrupt your mind."

Sundkvist's first project was to design and make a complete new set of shop tools and equipment. His work proved so effective that Maxim recommended him for membership in the Engineers' Club, and gave him a set of application blanks. But since he had had no schooling, he did not fill out the papers, or ever apply.

Many times Sundkvist proved worthy of Maxim's faith in him, and in his engineering ability. The prime example of this was his discovery of the value of fiber glass in the reduction of sound. On this occasion, Maxim placed his hand on Sundkvist's shoulder and said, "Others have tried, but you did it."

The adventures of these two experimenters on at least one occasion led them into the medical field. A doctor whose patient needed some unusual treatment, got Maxim and Sundkvist to design and make an apparatus of tubes and valves. It sustained the patient's life for a number of months. This schematic of HPM's is still on file.

The relationship between Maxim and Sundkvist was both productive and pleasant. Sundkvist once saw Maxim walking around on Whitney Street and asked, "Do you think it's safe?"

Maxim answered, "It always has been."

To which Sundkvist replied, "If all men were like you, it would be a better world."

"Thank you for saying that," Maxim said, "and I know you mean it."

In those days chief engineer Maxim was at the silencer plant regularly from 8:15 to 5:00. He drove to work in a big Hudson, and later in a Packard. Bourne, who was company vice-president in charge of research, did not relish riding in the back seat of the Hudson because it made him violently ill.

In March of 1921, Maxim wrote a comedy, "Miss Animator." His own description of the characters is indicative of the HPM style: "... a soap manufacturer, all around pessimist; a slick insurance office manager, looks like a million dollars, but you know he's getting away with about $2500. a year; head of Anglo Insurance Company, eminently respectable and thoroughly uninteresting; a typical political office holder, a bar room graduate, powers of comprehension nil."

Clever satire that it was, the production never achieved the success and fame of "The Virgin Paradise." But Maxim was not discouraged, and continued to produce many manuscripts in his spare hours.

*   *   *

In May of 1921 Maxim furthered the cause of amateur radio by getting the news direct for the Hartford daily when the wire service became disrupted during a storm.

In June, Maxim made the main speech at the dedication of Brainerd Airfield, in Hartford. This was the municipal airport that he had envisioned.

By August of 1921, the activities of commercial radio had added to the problems of the amateurs so drastically that the American Radio Relay League decided to hold their first national convention. The chosen site was Chicago, and the opening date September 3. Twelve hundred amateurs came from practically every state in the union. Hiram Percy Maxim gave the opening address. Secretary of Commerce Herbert Hoover sent a radiogram:

"The Department of Commerce is, by the authority of Congress, the legal patron saint of the amateur wireless operators. Outside of its coldly legal relations the Department wished to be helpful in encouraging this important movement. I am asking Mr. Terrell, the head of our Radio Division, to go to Chicago to learn where the Department can be of service."

Many years later Hoover's own son was to become a ham operator, and a leading figure in the League.

*   *   *

September of 1929 was an active month for the Maxim family. They traveled to Portland, Maine, where they visited Longfellow's house. They then went on to Gettysburg for a tour of the Battlefield. Pictures show HPM nattily dressed for these excursions, and characteristically holding his eyeglasses in his right hand. Other photographs of this period show Percy holding a gun equipped with her father's famous silencer, and the plant secretary, Miss Powell, firing a gun fitted with the silencer. She is wearing the hip-length belted sweater of the day, plus a V-neck blouse and a long, full-pleated, large-plaid skirt.

By late fall there developed an outgrowth of the first national amateur convention so important that it was termed "the greatest sporting event in the history of science." This referred to the first transmitting of amateur signals across the Atlantic, a possibility Maxim had envisioned even before the War.

On November 15, 1921, Paul E. Godley, one of America's receiving experts, member of A. R. R. L.'s Advisory Technical Committee, and member of the Institute of Radio Engineers and Radio Club of America, sailed for England on the Aquitania. He was given a grand send-off at a testimonial banquet and ham-fest of old-timers.

Almost a month later he was royally feted in London, and after setting up his apparatus for preliminary tests, moved on to Scotland. There on bleak and foggy Androssan Moor, beside the sea, he erected a tent and set up his receiving station. By midnight of December 7th signals began coming in. It was beyond doubt an American ham booming in through the heavy static, and the signal was 1 AW... the call letters of Hiram Percy Maxim!

The Founding Father reported this later in a Washington speech: "When the appointed night and hour arrived for the tests to be made, a terrible storm arose and Godley, located in a tent, faced the most discouraging conditions that could be imagined. Cold and wet, but jealously protecting his instruments, he maintained his night-long watch and established dozens of contacts with his American brothers.... The tests were repeated every night for a week, and every amateur in America had his chance to get his signals overseas."

QST magazine covered the history-making episode with the announcement, "Oh Mr. Printer, how many exclamation points have you got? Trot 'em all out, as we're going to need them badly, because WE GOT ACROSS!!!!!!"

The ten cold and rainy days that Godley spent in his tent proved another important thing too. More than two-thirds of the amateur stations who got their messages across were using CW, rather than spark transmitters. This settled a long-standing technical argument, and consigned spark to the scrapheap.

It was also in December of 1921 that HPM recieved a letter from his Uncle Hudson, in England, written just after he had attended Maxim's movie, "The Virgin Paradise."

"I am mighty pleased, after seeing it in Dover ... I have never seen anything better as a motion picture play... there was not one dull moment in it. I hope you will do more of this kind of work, for which you have such a crowning genius."

*   *   *

December was a pleasant month for HPM's daughter Percy also. She won first place in a horse show. A photograph represents her, wavy hair brushed back from her forehead at earlobe length, receiving the blue ribbon and silver cup. Her mount was Ellison's Pride.

By 1922, America was becoming less innocent, and much more noisy. Radical speeches moved from park squares to the breakfast table. Young flappers and shieks found wonderful excuses in Freudian psychology. Family life had new problems in cigarettes, gin, car rides and flamboyant confession magazines. There was a general breakdown of tradition and taboos. People were singing "Runnin' Wild," and "Hot Lips." But they were also singing Gus Kahn's "My Buddy," and Joyce Kilmer's "Trees." This was an age of experiment, and of consequent discovery. Amateur radio was one of the wholesome things which flourished.

Secretary of Commerce Herbert Hoover, with a genuine personal interest in the art of ham radio, offered a cup to be competed for annually. This was to be awarded to the best all-around amateur station, with particular preference shown to home-made equipment and individual effort. This award pointed up the important new status of amateur radio in America.

By January of 1922, there were, in addition to all the amateurs, twelve hundred broadcasting stations, most of them with insufficient equipment and inexperienced operators. These broadcasters had acquired hundreds of thousands of listeners. These were not technically-minded people, but they did have both affluence and influence. Since they had paid the high prices of those days for their radio receivers, they grew resentful of the hams, saying that they interfered with reception, and curtailed their pleasure. Politicians all the way up to the national level were besieged by their constituents howling, "Those damned amateurs... They bust up my concerts... What can be done about it?"

Amateur radio was in danger again. The ARRL, under Hiram Percy Maxim, laid down a plan of publicity and cooperation. They tried to secure better public understanding of the situation, and inaugurated silent periods during the evening hours as a gesture of conciliation. Finally, in March of 1922, Secretary Hoover called the first National Radio Conference. Something had to be done. There was conflict between the amateur and the broadcaster, and between corporate and private interests. Arrayed on one side were five big corporations, the American Telephone and Telegraph Company, General Electric, Western Electric, Westinghouse Electric, and the Radio Corporation of America. On the opposing side were the representatives of the amateurs, Maxim, Godley, Stewart, Warner, and delegates of various clubs.

The government, as far as it could, was on the side of the amateur. Hoover, in a rebuttal of press propaganda against amateur radio, stated (from the stenographic record), "The whole sense of this conference has been to protect and encourage the amateur in every possible direction."

The result of this conference was that certain meters were assigned to amateurs and others to broadcasters, and the term was legally defined for the first time: "An amateur is one who operates a radio station, transmitting or receiving or both, without pay or commercial gain, merely for the personal interest or in connection with an organization of like interest."

Amateur privileges, instead of being curtailed, had been increased.

The first National Radio Conference, in creating a definite line of demarcation between amateur radio and radio broadcasting, forced the editors of QST and Maxim, to make a decision, along with all of the other radio magazines of the time. Since there were only thousands of amateurs compared to millions of radio listeners, the other publications turned to the commercial field as a matter of economic profit. But QST, under the direction of founder Maxim, stood staunch for the amateur. It was ever thus with him. He put it this way: "With the amateur, radio communication is a labor of love, whereas with the professional it is labor for a day's pay. There is a tremendous difference between the two. Monetary return doesn't count with the amateur. It is the whole thing with the professional. No sacrifice is too great for the amateur to make, if he can get his signals through, and the answer back. Sitting up all night, sacrificing pleasures in order to save a dollar for the purchase of better equipment, trying innumerable experiments, rebuilding his apparatus time on end, never giving up, are a religion with the amateur. This passionate intensity of purpose is really nothing short of sublime."

Maxim regularly used the editorial pages of QST to support the position of the amateur everywhere. Typical of this is the following: "Amateur radio is still amateur radio...not a snobbish group able to buy the best, nor a group of engineers who know everything and never make a technical mistake."

On April 13th the first Transpacific two-way amateur communication was effected between Maui and California. About this time also, Atlantic operators were able to copy Pacific operators direct for the first time, and British and

French communication with the United States was established. The coincidence of these dates was probably due to favorable atmospheric conditions.

On May 17th of that year Maxim made a speech to the Hartford Business Women's Club. Its most memorable quote comes from a newspaper account: "Radio... is a hard thing to explain to women."

Nineteen hundred twenty-two was the year that HPM became the first president of the Hartford Engineers' Club. In this capacity he lectured before the New York Electrical Society, and spoke to numerous radio conventions. "I read in a magazine that amateur radio was doomed," he said. "This doesn't look like it."

In an article for the Harvard Crimson, Maxim commented on the history of radio, and its prospects for the future. He foretold the era of television, with its picture transmission... "The printed page does not get across... the radio telephone can do all this, and how much more is a matter of conjecture."

Maxim wrote a handbook titled "Radio for You and Me." In it he refers to radio as, "Man's mysterious messenger." About the work itself he said, "It is not a text book. It is not even a hard book... nothing but a friendly chat between you and me."

Eventually HPM did write a textbook, called "This World is Ours." In it he introduces his ideas about the universe. He explains that to a baby of one year, the world consists of its mother and its dinner. At ten a child has a clear idea of home and those in it, and of his town, and towns, but "there are lots of things of which he does not dream." At fifteen years, Maxim stated, "the ideas have begun," and at eighteen "He says to himself, 'so this is the universe!' Not at all... there are millions like ours, each containing hundreds of thousands of millions of stars like our sun... it seems to be a great sublimity working smoothly... a mechanism operating according to fixed laws."

It was also at this time that Maxim wrote a paper called "When the Ice Age Comes Back." In it he theorizes that "history is to repeat itself. We are in for another Ice Age." His description is exciting, and a bit frightening. "In the first Ice Age, man was not here... there were no cities... should the Ice Age come back, Man's mightiest works would have the same status as so many ant hills. We have built our civilization on the premise that the surface of Old Mother Earth is as enduring as the rocks. The trouble is the rocks are not enduring." He goes on to say that perhaps thousands of years away, the winters will become colder and colder... the summers shorter and shorter... that Canadians will be forced to move South, into the United States, and that all forms of travel except by air will be difficult. He predicted an exodus from the northern United States, with civilization centered a thousand miles south of its present general location. After a few centuries, New York would be as Babylon, Carthage and Thebes. No doubt HPM got into a lot of exciting arguments over these ideas, back in the year 1922.

During this time, HPM's daughter Percy was attending a "finishing school," which her mother had selected, and about which she herself was less enthusiastic. One of her classmates was Katherine Hepburn.

It was during these early 1920's that HPM and his daughter enjoyed a wonderful companionship and understanding. Every winter they went on what they called "bats" to New York. There they went to the theater, to concerts, and out to dinners. They saw La Boheme at least ten times, and enjoyed Cyrano, and a great deal of Shakespeare. They once went backstage, after a performance of "Rigoletto," and visited with Frances Alda. Maxim's favorite musical number was the Intermezzo from "Cavalleria Rusticana." Much to the despair of his family, HPM, when dining out at the most fashionable gourmet places, would order roast beef and boiled potatoes.

In the summertime, father and daughter enjoyed another kind of experience together. They would go boating and swimming in the Farmington River, on trips in the family Franklin or Packard, and fishing and camping deep in the woods, or yachting on HPM's "expensive toy," the Moby Dick.

Snapshots of this period show Maxim standing in the sagging doorway of a crude and isolated cabin in Maine. There is

Maxim and son, proudly displaying a string of little fish. There is Maxim seated in a canoe strewn with fishing gear; one boot is partially unlaced, and he is grinding away with a movie camera. There is Maxim at the wheel of an open touring car, with a large and attentive dog for a passenger. There is Maxim in cap and slicker, knee-deep in lake water, adjusting the motor on a folding canvas canoe, and there is Maxim in white ducks, peering proudly from the cabin of his trim yacht.

There was one scene on the Moby Dick that fortunately was not photographed. That was the occasion when the sea was unusually rough and Maxim had gone into the galley for some bread and butter. He soon realized that he should have omitted the butter. Seasick, and lying on the planks, he had to try and fix a conked-out engine, while Percy held fast to his legs, so that he could reach down. There were often other complications too, such as rain, mosquitoes, bugs, and a soaking wet dog.

HPM and his daughter had a flair for names. Their rather dangerous little folding canvas canoe was called Squiddidles, his pet name for Percy. Their uncertain-ancestry dog was called Lochinvar, or Lochie for short.

Some memorable nights were spent with their boat anchored in a little cove, quiet under a deep starry sky. HPM contemplatively smoked his pipe, and because he believed it hypocritical for his own generation to enjoy pleasures which they denied to their offspring, Percy smoked too. Her father had bought her a tiny pipe for this meditative sharing. For hours they would sit and think about the distant stars, and wonder about the possibility of life on other planets, commenting on the order, and the Great Design. It was a unique father-daughter relationship, all of this sharing of primitive beauty, sophisticated culture, wit and cosmic dreams. Percy perpetuated this rapport in a school essay entitled "My Pal and I," dedicated to HPM.

These were the days of "heavy living" in the Whitney Street house, when Maxim operated his ham rig in the conversary, and his wife entertained at bridge. There was an airedale next door, and HPM enjoyed seeing how much ice cream this dog could eat. He also got a chuckle out of feeding Cherry Bounce to the chickens to see how they would stagger and gawk. The house had a big walnut front door; Maxim would fling it open when he came home from work, and call out "Jess!" Then he would take off for a swim with Percy in the Farmington River.

Mrs. Maxim was her husband's severest critic, sometimes admonishing, "You just can't say that!" She often likened HPM to a fox terrier, because he was so quick, alert and active, with great concentration, yet never aggressive or forcing. He had an intent gaze, and smiled with his eyes as much as his mouth. He saw humor in almost every situation, and was never pompous. At one time he grew a beard, but as his wife did not like it, he shaved it off.

Maxim was always taking trains here, there and everywhere. He did most of his writing at night. He had terrible trouble with migraine headaches, and took short naps after lunch on his office couch, and at home after dinner. He described his headaches as a half-blindness, "as if you are looking through shimmering water and wearing blinders."

During the early 1920's Maxim enjoyed many motor trips with his family. These were adventurous experiences in those days of poor roads, and disintegrating tires.

While Mrs. Maxim enjoyed these motor trips, she often manifested some of the traits of a typical woman backseat driver. Maxim gives a newspaper account of one such excursion in the rain: "The ride inspired Mrs. Maxim to flights of oratory... running down the long slippery grade on the top of a narrow embankment causes her to press toward the high side and utter hissing sounds... she grows remonstrative, refers to our past life, expresses certain statements regarding the hereafter... and while possibly not downright uncomplimentary, she lets it be very plain that she does not approve of me or my driving."

It was about this time that HPM originated the Aero Club of Hartford, where he was for many years chairman of that city's Aviation Commission. He was also chairman of the Hartford Branch of the American Society of Mechanical Engineers, and chairman of the Connecticut Section of the Society of Automotive Engineers.

Things were progressing well at the silencer plant. Cecil Powell was HPM's secretary and alter ego, ever ready with practical help. She also experienced some of the practical jokes which her boss and Bourne were famous for. The latter once planted some corn in the flower pot on her window sill. It got up quite high and she hadn't noticed it, but HPM did. He asked her about it, saying, "It looks like corn to me."

When she replied that it couldn't be, HPM looked Bourne right in the eye and said, "It's Golden Bantam."

In those days Sundkvist had to make many decisions, rather than delegating them. But he always listened to HPM, particularly in financial matters. "Five hundred dollars per share for good stock is better than five hundred shares of poor stock at one dollar," Maxim said. "Why isn't the best good enough?" Perhaps his worst business failing was that he believed everyone was as honest as he was.

There are some pleasant home movies of this period. They show HPM as a scientist, inventor and author, at home and at work... intently studying and patiently observing. They show him taking off at Brainerd Field in an open cockpit biplane; presiding at the American Radio Relay League, and at the Cinema Club... eager, inquiring, responsive, with finger tips and thumbs together, wearing a bow tie, and eyeglasses on a black cord. They show him on the Moby Dick, wearing his skipper's cap, and clamping a pipe in his mouth. They show him conversing in his home with a friend... a rapid speaker gesturing with a finger, smoking, and peering into the flames of the fireplace... "out of the Cosmos, as it may be revealed to him."

By the year of 1923, amateur radio was on the threshold of its greatest development, the exploration and use of the short waves. This was made possible by the wide use of the vacuum tube instead of spark, although a few of the Old Guard still flaunted "Spark Forever" on their QSL (acknowledgment) cards.

In October, the Third National Radio Conference took

place in Washington D.C. The Secretary of Commerce gave the keynote of this meeting in his opening address:

"Nor have we overlooked in these previous conferences the voice or interest in the amateur, embracing as he does that most beloved party in the United States ... the American boy. He is represented at this conference, and we must have a peculiar affection for his rights and interests. I know nothing that has contributed more to sane joy and definite instruction than has radio. Through it the American boy today knows more about electricity and its usefulness than all of the grown-ups of the last generation."

At this conference a subcommittee was appointed to study amateur problems. Their chairman was Hiram Percy Maxim. Some of the new difficulties had international ramifications. Whereas in the beginning, foreign experimenters regarded American operators as something akin to gods, they began to lose some of this respect as their contacts became more numerous. The British in particular were disturbed by the brash manners of some of the American youths, and the abruptness, partly due to the code, of many of the oldsters. This situation never became critical, however, and overseas operators soon accustomed themselves to America's casual social standards.

With radio communications flourishing on several continents, the endeavor took on such an international aspect that Esperanto was adopted as the official language. However, it never came into popular use with the amateurs. Instead, they developed their own international language, referred to as "QST English." It is a combination of abbreviations of English words and code signals, and it became an effective kind of radio shorthand all over the world. Hams everywhere know that "73" means best regards, and that "88" is love and kisses. "C" signifies yes, and "N" no. A QSO is a communication; a YL is a young lady; and XYL is a married woman; TNX is thanks, and so on, into a long list of symbols.

It was in this year of 1923 that amateur radio had an opportunity to prove its worth on an expedition. In the spring,

Captain Donald B. MacMillan, who had already made eight journeys into the Arctic regions, was organizing a bigger and better one. Having previously experienced the near-disastrous effects of prolonged isolation, he went to Hartford to interview Hiram Percy Maxim on the possibility of using amateur radio. While a guest in the Maxim home, they worked out an arrangement together, and when MacMillan's Arctic Expedition set sail from Wiscasset, Maine on June 23, it was provided with communication facilities. The little auxiliary schooner Bowdoin had been fitted out with a complete 200 meter amateur radio station donated by the Zenith Corporation, and furnished with an amateur operator, Donald H. Mix from the ARRL.

Besides standing his regular watch, Mix, during the following months, transmitted a weekly 500-word message to the North American Newspaper Alliance, handled amateur traffic, and kept a record of amateur calls heard. He flashed the news back to civilization when the expedition erected a National Geographic Society bronze memorial to the Greeley Expedition on the spot where it had perished from starvation and exposure. Mix's "Wireless North Pole" set a new DX (distance) record.

That little station on the Bowdoin was a great blessing to the expeditionary party. It did away with its isolation, provided entertainment, and relayed news of the outside world. The men learned the result of the Dempsey-Firpo fight, and all about the terrific earthquake in Japan. That Christmas, President Coolidge sent season's greetings to the party, which was frozen in for the winter in Greenland.

When Mix returned with the crew the following September, it was to a new world of amateur radio. His 200 meters had become obsolete through the use of the short waves. But he had established himself in history, and Captain MacMillan said that "No Polar expedition will attempt to go North again without radio equipment." This proved to be the literal truth.

Also at this time, through the generally accepted use of the new vacuum tubes, it became possible for the amateur to transmit the human voice, in addition to the code signals. This was a high point of expansion for hams. When they became vocal, they had a whole new field of human contacts, complete with "any third parties" who might be ham shack guests.

*   *   *

The year 1924 saw America blessed with the melodies of Berlin, Kahn, Gershwin, and Hammerstein, and with a great new favorite, "Tea for Two." It also saw HPM and Bourne in the back seat of a New York taxi, where they drew up plans for an industrial silencer for American Telephone and Telegraph. The company used it on their first testing engines, all over the United States.

Maxim, a leader in the amateur cinema movement, took many reels of movies during this period. One of them shows him with Eastman and Edison, the discoverer of film, and the inventor of motion pictures, respectively. Other subjects include General Pershing; Dr. H. E. Ives of the Bell Telephone Laboratories; Adolph Ochs, dean of American journalism; E. A. Alexanderson, pioneer of radio and television; Dr. W. D. Coolidge of General Electric; Dr. G. K. Burgess of the U. S. Bureau of Standards; James Harbord, president of the Radio Corporation of America; Canada's Owen D. Young; and David Lawrence. These scenes show Maxim's twinkling eyes, thick hair, and eager forceful manner. He has the familiar glasses on a neck cord, and the characteristic manner of throwing his head back when he laughed.

Other movie scenes of this period show Maxim crossing the Delaware at Wilmington, and a night view of Trenton, complete with theater signs. They include shots of Mrs. Maxim driving through city traffic, with its street cars and vintage vehicles, and of the Maxim home, with its bird bath and garden, dominated by a very high radio pole. Other reels are of scenes from Maxim's office window in Hartford, showing a stream of moving vehicles in the intersection below, with a policeman directing traffic.

In contrast to the machine gun and ammunition contributions of the other Maxims, HPM was a man of peace. His greatest hope for amateur radio was that it might be the means of uniting all the continents of the world in lasting peace. To this end, early in 1924, the American Radio Relay League asked their president, HPM, to represent the League at a meeting in Europe. Accordingly, on March 12, 1924, at the Hotel Lutetia in Paris, a historic session was held. The radio representatives of nine nations gave a dinner for Mr. Maxim. Among them were some of the most distinguished radio men in Europe. The countries represented were France, Great Britain, Belgium, Switzerland, Italy, Spain, Luxembourg, Canada, and the United States. Mrs. Maxim acted as interpreter. At this meeting a new organization was formed, the International Amateur Radio Union. Maxim was made chairman of the temporary committee to effect a permanent organization, and draft recommendations and a constitution to be presented at a meeting during the Easter holidays of 1925. HPM expressed his peace aims in the closing words of his address: "The spirit of brotherhood that pervades amateur radio in America can be made to spread its mantle of good fellowship over all the civilized nations of the earth."

Back home, Maxim gave an enthusiastic account of the historic conclave in the May 1924 pages of QST:

"This ARRL President has sat in at a good many very impressive radio meetings in the past, ranging from Maine to California, but he has never sat in at a meeting where there was quite as much thrill as this meeting in Paris where the amateurs of nine different countries sat down together.

"The chairman of the Inter-Society Committee of the three important French radio societies, Dr. Pierre Corret, presided. After the dinner he opened the meeting with a stirring address delivered in French... in the most dramatic fashion Dr. Corret pointed to the great amateur society across the ocean which the world looked upon as a leader in not only the science of radio communication, but in the humanitarian aspects of radio. He paid, with graceful eloquence, the compliments of Europe to the President of the ARRL [Maxim] and certainly

placed it up to the latter to come through as he had
never been required to come through before.

"On such an occasion ponderous oratory seems
out of place, and recourse was had to the simple
American method of stating the case as exactly and
briefly as possible, and asking for sincere consid-
eration... American amateurs wished only to suggest, and under no circumstances to be considered
as dictators.

"It would be difficult to describe the atmosphere
that pervaded that dining hall at the termination of
these two addresses. Those of us who have attended American 'hamfests' have some idea of the thrill
that went with this thing... I do not believe I am
overstating the facts when I say that every man in
that room will remember distinctly all his life the
keen thrill of that meeting in the Hotel Lutetia in
Paris that evening of March 12, 1924."

On June 14, 1924, Maxim was invited to present the
Commencement address at Colgate University. On this occasion he was given an Honorary Doctor's degree. As might be
surmised, the subject of his talk was radio: "Radio is one of
the most wonderful scientific achievements of man." Of Marconi he said, "... of course there were those who laughed.
They said as much could be accomplished by means of a lantern... an ordinary kerosene oil lantern. I suppose we shall
always have these shortsighted people among us... Marconi's
5 miles soon became 30... beyond the field of the oil lantern
... and the 30 became 100, and ships at sea could communicate with shore." He spoke of the World War, and his atten-
dance at a conference at the Brooklyn Navy Yard to get help
from the telegraphers. Then he said, "... now comes the international amateur... meeting in Paris next year [1925].
Next is outer space." He concluded with this statement, "The
outstanding feature in radio influence is exerted by young men
... almost always poor boys."

In the winter of 1924, amateur radio had another chance

to prove itself. Half of the United States had been paralyzed by a blizzard, and a terrible sleet storm. Practically all of the communication wires in the North and Middlewest were down. Cities were cut off, and lives and property were threatened. Amateur operators rallied to help. A Minneapolis station did a Paul Revere, with amateur Minutemen. DesMoines helped a disabled mail plane, and transmitted news of the arrangements for Wilson's funeral. Nurses, doctors, and first aid were provided. Even the railroads utilized the services of the hams.

*   *   *

By 1925, the whole United States seemed to have become casually unrestrained. Pola Negri and Gloria Swanson enthralled the men, while women by the thousands fawned over Rudolph Valentino. We had Mah Jongg. We had speakeasies. We were beginning to pile up automobile death statistics. Clara Bow had developed into the "It Girl." Cecil B. DeMille was specializing in opulent movies. We inaugurated national bathing beauty contests, and discovered Main Street. This was a year of wonderful songs, including "Moonlight and Roses," and "Always."

It was a promising year for hams also. They were too busy with antenna, tubes and schematics to even realize that there were such things as behavior problems, frustrations and complexes. They were working the Antipodes directly. The once-despised waves between fifty and two hundred meters were performing miraculously, spanning the earth every hour of the day and night. The greatest possible terrestrial distances had been conquered.

In January of 1925 Percy Maxim, scorning the idea of a debut, announced her engagement to John Glessner Lee, who had been a classmate of her brother's at MIT. The newspaper story commented: "His paternal grandfather was Major General Stephen Lee of the Confederate Army, later president of the University of Mississippi. Miss Maxim comes from a long line of inventors and engineers. Her father, Hiram Percy Maxim, inventor of the Maxim Silencer, and president of the

Maxim Silencer Company, is an engineer of high standing, and has done considerable work in radio and research development. Her mother, once Josephine Hamilton, is the daughter of Governor William Thomas Hamilton of Maryland. Her other grandfather, Sir Hiram Stevens Maxim, was inventor of the machine gun bearing his name... No date has been set for the wedding, for Miss Maxim is but 19 years old, and must be six months yet in Miss Master's school at Dobbie Ferry before her graduation."

The engagement announcement was printed in the January 4, 1925 issue of Vogue Magazine.

Percy's class photograph, taken during this period, shows her with her hair shorter and smoother, with a high left part, and brushed softly across her forehead to the right. It is an appealing face, serious and purposeful. Her dress had a simple V-neck collar.

A 1925 picture of John Lee presents him as being young and handsome, with thick dark hair, and wearing no-rim glasses. His expression is serious, but with a distinct hint of humor. For the first time in his life, HPM's attitude became thoroughly disagreeable, almost discontented. He did not want to face up to the idea of his daughter's marriage, because through it he would lose his companion.

Fortunately, he was distracted by the opening of the scheduled meeting of the First International Amateur Radio Congress, which took place on April 14, 1925, in Paris. Mrs. Maxim accompanied her husband on this occasion, and acted as official interpreter for the group. Four days later the details of organization were completed, and Hiram Percy Maxim was elected international president. A total of twenty-five nations was represented: Argentina, Austria, Belgium, Brazil, Canada, Czechoslovakia, Denmark, France, Finland, Germany, Great Britain, Hungary, Italy, Japan, Luxembourg, Netherlands, Newfoundland, Poland, Spain, Sweden, Switzerland, Uruguay, the United States, and two late arrivals, Indo-China, and the U.S.S.R.

There is a complete motion picture record of this trip, beginning with the Mauretania in the East River just before sailing. There are shots of the crew, and of a crowd waving from the pier as the ship slowly pulls away. There are glimpses of Warner walking on deck in a strong wind, of a game of deck tennis, and of a fast liner on the Mauritania's bow wave.

By Friday afternoon they were off Cherbourg, and the film shows a tender coming in to take "us Paris passengers off." On the sixth day from New York the delegates were in Paris. There is a movie night scene, in the rain, showing the Arch of Triumph, France's Tomb of the Unknown Soldier, and Warner at the base of the Eiffel Tower, dapper with a cane.

Between radio sessions, the party extended their European tour. They spent Easter in Normandy in a chateau on an old French country estate. Film glimpses here show Mrs. Maxim wearing a long suit with a three-quarter coat, blouse, fringed scarf, and pearl beads. HPM is meditative, with his hands clasped together. Quite obviously he was enjoying this trip. This is shown particularly in the Park scenes, with their shots of lovely French children, fancily dressed, briskly walking, and of handsome, neat little boys sailing toy boats. These charming movies continue at the Champs d'Elysee and the Place de la Concorde. At Mr. Waddington's den in Paris, a sort of "roundhouse of stone," the camera focuses on the date "1162" over the door.

After a week of radio activity, and a banquet in their honor, the Maxims said goodbye to Paris and boarded the liner at Calais for the return trip. Maxim wrote, "It was the roughest crossing the Channel I ever made... almost impossible to hold the camera." Maxim took pills, but still suffered from seasickness. However, he recovered enough to take many reels of pictures.

In May of 1925 the Maxims acquired their summer place, Bill Hill, in Hamburg, near Lynne, Connecticut. It had been built in 1760, and still had the heavy board shutters with which it had been fitted as a protection against Indian attacks. The area on which the house stood had been farmed by a family by the name of Bill, who had named their sons after Congressional Acts. There was Louisiana Purchase Bill, Missouri Compromise Bill, and Kansas-Nebraska Bill. The place was very rural and remote, offering real escape from the city. There was an old stone fireplace, and much of the original window glass. There were five tumble-down old barns, and a mountain of scrap metal. The Maxim family personally undertook the cleaning-up operations, and recorded the progress on movie film. There are shots of them hauling away an ancient kitchen stove, tied onto the side of an old Essex. They removed tons of rusty iron, raked the huge yard, and inspected the well, with its pulley and cable. The weather was cold, and Mrs. Maxim is photographed watching the operations, wearing furs and galoshes. There is a humorous sequence of HPM and his son tying a rope around the old three-holer and dragging it away. Gradually their retreat was becoming habitable.

Maxim had dropped the manufacturing of gun silencers, and was concentrating on industry operations. The Diesel engine and the internal combustion engine had increased his business greatly, and he was prepared to silence any noise that came out of a pipe. It was about this time that Bourne made a junket to Virginia for the firm, with briefcase and slide rule, to silence a silk mill that could be heard for miles. He ate a huge Southern breakfast, then figured for two hours with his slide rule to determine what was wrong with the installation. He changed the pipe accordingly, and although he later found that the calculation was totally wrong, it worked!

During this year of 1925, friendly relations had been established between the armed forces of the United States and the radio amateur. The Navy contacted the ARRL, and using one of their operators plus his apparatus on the flagship Seattle, proved the efficiency of the short waves. Shortly after that, on June 19, Hiram Percy Maxim was commissioned a Lieutenant Commander in the United States Naval Reserve.

In July the Maxim family took up their summer residence at Bill Hill. The old well lasted three days. The only permanent solution seemed to be an Artesian well, and there are movies of a team of horses dragging in a steam well rig. The cost was ten cents for each fall of the beam, and the drilling went on for two and a half weeks! The camera recorded the sharpening of the bit, its end white-hot, with a Model T in the background. HPM is shown too, using a shovel, for the digging of post holes. Outside shots include luxuriant vines, flowers, and a picturesque stone wall. Inside views show the fireplace, cupboard, and rustic furnishings. It was a bonafide 18th century house, made pleasant and habitable.

In August the Army sponsored the formation of the Army-Amateur Radio System. This affiliation was successful, and was featured in official maneuvers.

*　　　　　*　　　　　*

The year 1926 saw "The Birth of the Blues," and an exciting motion picture, "The Desert Song." Maxim fitted out the third floor of his silencer factory to make radio equipment. Tuska went on to form the Tuska Radio Company. Business proved to be good, but there were patent problems. Bourne was the company engineer. He became an expert in radio accoustics, and was made vice-president in charge of research. Bourne's father had been a minister, in very moderate circumstances, but a wealthy railroad man, recognizing the young man's potential, had financed his technical education.

This was the year of a big social event in the Maxim family. The Hartford newspaper told it this way:

"June, the month of roses and brides, will be ushered in Tuesday by one of the most important weddings of the season. Miss Maxim but recently returned from Europe... "

The date was June first, the place the Maxim residence on North Whitney Street, at 2:30 P.M., for the immediate family, with a reception afterward. Percy's brother Hamilton was best man, and John Lee's sister was maid of honor. The house was decorated with blue and white cut flowers, plants, ferns and palms. Reverend William Hooper, rector of St. John's Episcopal Church, officiated, and the bride was duly "given in marriage by her father." She wore a white chiffon gown of youthful style, and a veil of white tulle held in place with a simple band of Brussels lace. She carried a shower bouquet of roses and lilies of the valley. An orchestra played for the reception in the garden, which had been decorated by members of the Hartford Garden Club. The couple left in their roadster, a gift of the bride's parents, for Detroit, where John Lee was to be an aeronautical engineer with the Aeroplane Division of the Ford Motor Company.

Shortly after this event, Maxim and his wife left on a European tour.

Movies made later that summer show the Maxims and the good ship Seagull, twenty-eight feet long, and able to go anywhere. The water was cold and rough, off the Connecticut shore. HPM was at the wheel. Maxim's son Ham had a small boat which had cost only a little over two hundred dollars, and was equipped with a Johnson motor. They went to watch a yacht race, with Mrs. Maxim aboard. She did not care for it, but was a good sport. HPM, wearing a high stiff collar, tie and hat, took movies of the competing yachts on their way to Bermuda, a voyage of four or five days. One yacht, the Julie Briese, had sailed the Atlantic to enter this race.

Maxim enjoyed photography. He had a big crank camera on a tripod, and was constantly experimenting with new gadgets, including color apparatus. In 1926 this included a red, blue and green glass filter over the lens, and required the same process when projected. Maxim developed a prismatic attachment that permitted the camera operator to shoot pictures at right angles, after the manner of a submarine periscope. He had a great deal of fun experimenting with this in his sun parlor, ostensibly aiming at one person, but "shooting" another. It was in this year, 1926, that Maxim founded the Amateur Cinema League.

The summer of 1926, to the consternation of the hams, had seen the complete breakdown of radio law and order. The Hoover Radio Conferences, designed originally for recommending Congressional action, had turned instead into sessions in which the radio interests mutually agreed to observe the regulations which had been set up by the Department of Commerce. This gentleman's agreement had served adequately until the decision was handed down in the Zenith case on April 16, 1926. It was made public by the Attorney General some weeks afterward that in the eyes of the law, the federal government had no control over radio excepting as authorized by the 1912 Act. This of course had included no reference to broadcasting, or to high-frequency allocations, since they were unheard of at that time.

When this situation was announced, hundreds of broadcasting stations immediately jumped the frequencies to which they had been assigned, in favor of better spots. Moreover, they stepped up their power as they saw fit, regardless of causing interference. With many new stations coming onto the air, the effect was chaotic. The radio amateurs did not join in this stampede, however, but stayed within their assigned bounds. It was a wonderfully effective act of self-discipline and self-regulation, supported by Maxim's editorials in QST magazine.

The ARRL had other problems on its mind in 1926, and they involved other countries. The International Broadcasting Union insisted that amateurs be assigned very low power, and certain narrow bands, so as not to interfere with broadcasting. The ARRL prepared the amateur's position, and presented his case successfully.

The year 1926 saw amateur radio scoring again on numerous expeditions. This was when Lieutenant Commander Richard E. Byrd sailed with his Byrd Arctic Expedition for Spitzenbergen, assisted by short wave equipment and amateur operators. The American Museum of Natural History sent an expedition to Greenland, relying on amateur radio for contact. In the fall, the Chicago Field Museum and the Chicago Daily News sent an expedition to Abyssinia, with short wave radio equipment and an amateur operator. About this time too, the Roosevelt Memorial Expedition set out for the wilds of Brazil, also with portable amateur equipment. In addition to these major explorations, there was widespread use of the short waves by ships and yachts.

*　　　　　*　　　　　*

In 1927 the country was singing forty-four great hit par-

ade songs, including "Girl of My Dreams," "Blue Skies," and
"Hallelujah." There were many spectacular movie musicals,
including "Showboat." At this time amateur radio was to make
great progress too, because of the enactment, in February, of
the Radio Act of 1927. This provided for the creation of a Federal Radio Commission, to have control over all radio matters. The ARRL besought President Coolidge to appoint some
of their number on the five-man commission. Of the names
submitted, only one candidate, Colonel John F. Dillon, was
successful. However, the amateurs were treated well by the
Commission. Legislative action had become imperative because of the hopelessness in which the broadcasters had en-
tangled themselves. The Federal Radio Commission was giv-
en power to classify radio stations, to assign frequencies, al-
lot power, determine location, regulate apparatus, and to re-
quire logs. Violation of this Act was deemed a criminal offense. The Secretary of Commerce was empowered to inspect
stations, discipline operators, and to assign call letters. It
was in this Act of 1927 that the word "amateur" was used for
the first time in a statute.

During the summer of 1927, the Maxim family enjoyed
its usual pursuits. Movies of this time show HPM yachting,
cruising in the Black Duck, or with the Sea Gull towing a din-
ghy. He is wearing white ducks, even though it was impossible to keep them clean. The Black Duck was Ham's first
boat, and had an antique, undependable engine which made
calm weather most welcome.

On May 6, 1927, Hudson Maxim died. In the latter years
of his life he had been occupied with experiments in color photography. Minus one hand from an explosion, and trying to re-
strain himself from the fifteen minute bouts of swearing which
were the despair of his church-going wife, he presented movie
demonstrations of film which portrayed shades of red, green
and beige. He was a successful pioneer in this field.

In the summer of 1927, amateur radio was employed to
provide communications for the Dole Air Races between California and Hawaii. It functioned well, but its most memorable
reporting concerned two missing planes, the Miss Doran and

the Golden Eagle, and the gallant but tragic search made for them by Captain William Erwin and Alvin Eichwaldt. These two amateur operators, flying in the Dallas Spirit, kept their signals coming in steadily in spite of the distance and the bad weather. They lost speed and went into a tailspin, with Eichwaldt still at his post, sending calmly. Their signal rose to a shrill shriek and then fell to practically zero with the violent pitching of the plane. Finally the dots and dashes fizzled out as their antenna hit the water. Alvin Eichwaldt, knowing that he was headed for certain death, had stayed at his key to let the world know what had happened. This was in keeping with the tradition of the amateur radio operator.

Of great importance to all hams was the International Radiotelegraph Conference, held in Washington D.C. in the late fall of 1927. There were 351 delegates from seventy-four nations and fifty associations. The group was addressed by President Coolidge and Secretary of Commerce Hoover, who was president of the conference. The unwieldy body was divided into committees and sub-committees. The amateur was represented most ably by Hiram Percy Maxim, of the ARRL. His problem was that most of the other countries of the world had no conception of the fact that an amateur was ever anything but a liability, cluttering up the airwaves. These countries seemed incapable of believing that our government granted the amateur his privileges of its own free will, rather than under the stress of political pressures, and they were afraid of what such an action might mean in their countries.

As the conference dragged on, delegates in small groups discussed matters over innumerable cups of tea. When a few men finally hit upon a point of agreement, they would bring others into their discussion circle. The amateur was supported in this "tea cupping" by the entire American delegation, from Hoover on down. They held out for all they could, but some compromise was necessitated in order to come to an agreement among all the nations. Maxim and Charles Stewart (vice-president of ARRL, and vice-president of the International Amateur Radio Union), believed at this point that discretion was the better part of action. Stewart had been gambling to gain more, but they compromised and held out for 400 kilocycles at 40 meters. As a result of these treaty provisions, the American amateur lost over thirty-seven percent of his territory. Of course the reverse was true with the other nations, who found themselves with greatly increased privileges.

Concerning this episode, Maxim wrote an editorial for the September 1927 issue of QST. It was entitled, "The Reason Why":

"Sitting back in the old armchair, with the last issue of QST read from cover to cover and with everybody else in the house asleep hours ago, I fell to thinking of amateur radio today and amateur radio of other days. As the blue smoke curls slowly upward from the old pipe, visions of early ARRL Directors' meetings float before me. I see those oldtimers grappling with problems of organization, with QRM [frequency noise and interference], with trunk-line traffic and rival amateur leagues. I see sinister commercial and government interests at work seeking to exterminate amateur radio. They were dark days, those early ones.

"Today I see amateur radio an institution, recognized by our American government, and on the road to recognition by the other great governments of the world... I see a rapidly-developing worldwide amateur radio brotherhood taking shape, in the form of our I.A.R.U.

"And as the last embers of the old pipe slowly turn to grey ash, I ask how it all came about: that the ARRL should have succeeded and all its opponents failed. The answer is clear. It is because with our opponents there was always some kind of a selfish motive to be served for someone, whereas in our ARRL we insisted from the beginning that no selfish motive for anybody or anything should ever prevail. Everything that our ARRL undertakes must be one hundred percent for the general good. That policy bred loyalty and confidence. With those two things an organization can prosper forever."

*   *   *

Nineteen hundred twenty-eight was another year of catchy melodies, including "Shortnin' Bread," and "Carolina Moon." Al Jolson was singing "Sonny Boy," and Romberg and Hammerstein had promoted "Stout-Hearted Men." In February, 1928, Carl Laemmle, president of Universal Pictures, wrote a letter to Hiram Percy Maxim, asking for his autograph. He enclosed a special parchment sheet for the purpose, and explained that the signature was for Laemmle's own personal album, which he had kept since he was seventeen.

In 1928 Mrs. Maxim was elected to the Hartford Board of Education, and was serving a second term on the Hartford Board of Health.

On May 15 of that year, Maxim was appointed as Aviation Commissioner for the city of Hartford. He had also made great advancements in his silencer business. Some of the machine units were over six feet in diameter. And HPM's newest development was so large that only one could be loaded onto a railroad car. These were called Maxim Evaporators, a new experiment trying, at low cost, to make fresh water out of salt water.

On October 9, 1928, Maxim made a speech at the Hartford Steam Boiler Insurance Company. His topic was the gas engine, over which, as his notes state, "more profanity has been evoked...!" His remarks concluded: "I have always felt that it was foreordained in the late 1840's by Samuel Kier when he concluded that beeswax, tallow fat and whale oil did not make a completely satisfactory illuminant... of all the cranky, tempersome, unreasonable, impossible pieces of machinery that the brain of man has devised, the gasoline engine is the most."

In that year of 1928, HPM and Josephine took a Mediterranean cruise on the Berengaria. Maxim took many reels of pictures. These he developed on board the liner, as a friend put it, "because on board he had nothing to do, since he couldn't run the ship." HPM took dozens of shots of the Pyramids, and other "tourist attractions," despite the jibes of his friends, who pointed out that you could buy much better ones with a fraction of the effort and expense.

*　　　　*　　　　*

By the year 1929 motion pictures had begun to talk, and the hit songs ranged the gamut of moods from "Great Day" to "Am I Blue?" America's economic tailspin had begun, and the Maxim business interests went down also. In June Josephine suffered a mild stroke, which added to HPM's worries and financial stress. Percy, married, with a family, and living on Long Island, was not able to help much, but she performed such services as she could. The Depression was awful, and forced a really tough squeak for the Silencer Company.

It was about this time that a minister asked Maxim, who was not noted for his church-going, to come and preach the Sunday sermon. HPM declined, but said if he did, it would be a good one.

It was also during this period that someone pointed to Maxim on the street and asked his companion, "Do you know who that man is?"

"Yes," the answer flashed. "That's Hiram Percy Maxim. He'd rather experiment than eat."

In 1929 a new plan for the Army-Amateur Radio System was put into effect. Its purpose was primarily to aid the Army and the American Red Cross to help distressed areas in times of emergency and disaster.

As a demonstration and test of amateur ability, another Governor-President Relay was conducted, with forty-one official messages sent to President Hoover. That gentleman was already conversant with the problems and aspirations of amateur radio through the experiences of his son, Herbert Junior, who was a licensed amateur and a member of the Washington Radio Club.

It was in this year of 1929 that the first conviction was obtained under the Radio Act of 1927. It was against an unlicensed operator in St. Louis. The unfortunate fellow was sentenced to a year and a day in Leavenworth, but instead, was deported as an alien. This affair proved that the new law had teeth, and was evidence to the amateur that the regulations meant what they said, and were to be strictly heeded.

Early in January of 1930 a bill was introduced in the United States Senate by Senator Couzens of Michigan, to create a national communications commission to control all forms of wireless communication. Hiram Percy Maxim was requested by the American Radio Relay League to go to Washington to testify before the Interstate Commerce Committee as to the value of the amateur to society, and the desirability of considering his point of view in any contemplated legislation. It was an instructive and moving speech, the highlights of which were as follows:

"As radio spread into wider popularity, and new channels became necessary for commercial users, the amateur was made to give up part of the territory he had pioneered. This did not sit comfortably, especially since his numbers had grown to many thousands in the United States alone. More and more continued to be taken from him until the International Radio Conference in 1927, when he was all but sacrificed by a conference of delegates from some eighty countries, most of which had no radio amateurs, and where every form of communication is a State monopoly. Our American delegates from our Army and Navy and Department of Commerce fought valiantly for their American amateurs, but they were in the hopeless minority. All that was left of the territory the amateur had so brilliantly chiseled out... were extremely narrow bands of frequencies... here we find him today [1930]... crowded beyond all conception and almost beyond endurance, suffering from what he feels is injustice and ingratitude, and in constant danger of losing even the little that he retains."

HPM concluded: "However, any governmental regulative agency will be besieged by commercial interests to grant even greater numbers of channels to them. It will be urged upon such an agency that the amateur channels are worth thousands of dollars in earning ability. Let me entreat that you believe that those few radio channels which have been left for the amateur constitute a value to our nation incalculably greater than any possible money earnings that could conceivably be developed from them by any commercial company.

"Finally, in considering means of regulating communications, let me urge you with all the force and sincerity I can command that you bear in mind the radio amateur and make possible his continuance. To do less would be nothing short of national catastrophe."

The bill failed to pass, and the basic radio laws were not changed. Hiram Percy Maxim's speech went down in communications history as one of the strongest documents ever written on behalf of amateur radio.

Maxim was a rare traveling companion, and his associates on these occasions, such as this Washington trip, had many vivid memories of their Chief. In the early morning he indulged in various setting-up exercises in his hotel room, bending over so far that he could touch the palms of his hands on the floor. Then he would proceed to gallop about the hotel room in his pajamas, on all fours. He always preferred a cold bath. With the tub full of icy water, he would perch on its side and then slide down the inclined end in a tremendous splashing, and flailing of arms and legs, yelling at the top of his voice.

HPM's firm had been developing the Maxim Window Silencer, and the October 23, 1930 issue of the Hartford Courant carried a headline: "Maxim device to bar noise from rooms is successfully tested... a boon to hospitals and office buildings." It elaborated, "to windows that open on a noisy corner it gives perfect quiet."

A photograph of Maxim, showing his inquiring eyes with the laugh wrinkles at the corners, was featured, with the caption, "Maxim adds to scientific gifts to humanity."

Quoting from the October 25, 1920 issue of the New York Times:

"Think of Hiram Percy Maxim walking along the noisiest city in the world with all the sound rippling away from him... He has silenced grinding machines and crackling rifles... muffled the roar of the Diesel engine... his new device combined a ventilating fan with a silencer... a boon to homes near construction sites, and best of all, the individual who retires early can open his window, adjust the silencer, and sleep without hearing the blast of brasses from the neighbor's radio."

This was the prelude to the development of air conditioning.

In December of 1920 the Hartford newspaper carried an item of amateur interest: "Maxim to tap key to open distant show... The Australian Radio Exposition begins Monday, as the flash is sent at Western Union here. Hiram Percy Maxim, president of ARRL and of the International AR Union, officially opens the Exposition in Town Hall... Melbourne, Australia, 7:30 P.M. Australian time." The signal exploded a tray of flashlight powder to take the photograph of the Australian opening, through the cooperation of Western Union, and Canadian and Australian telegraph companies.

In September of 1931 the Hartford newspaper carried a picture of HPM, stating "... Hiram Percy Maxim, whose father and grandfather turned their inventive genius to destruction, was celebrating his sixty-second birthday today in rejoicing at the perfection of his latest device to make the world more peaceful. 'A device for you, and you and you,' he says ... a plain metal box which keeps out all noise while permitting fresh air to enter.'"

This pioneer version of an air conditioning unit was called by Maxim "the result of twelve years of work."

On October 1, 1931 a letter from the Office of the Chief

of Operations of the United States Naval Department was sent
to HPM requesting his photograph, to be hung officially along-
side those of Marconi, Bell, Faraday and Morse. It was further asked that it be autographed.

Two months later HPM received the following invitation:

The President of the United States
invites
Mr. Hiram Percy Maxim

To attend the meeting of the President's
Conference on Home Building and Home Ownership

Which is called in the city of Washington
December second to fifth
1931

Later in December of 1931, QST magazine carried an
editorial on the origin of the word 'ham.' It had nothing to do
with a similar word in an old song, nor had it any reference
to the theatre's "Hamlet." Rather it was derived, according
to the account, from the British Cockney version of amateur
(h'amateur), abbreviated into 'ham.' "And," the editorial concluded, "Hams we are, and proud of it."

By the year 1932, America was deep in the Depression.
We had experienced drouth, the Scopes trial, Lindberg's flight,
and Byrd's ham radio reports from Little America. Amos and
Andy were nightly broadcast fare. The world had come to know
Walt Disney, Babe Ruth, Will Rogers and Man O' War. Veter-
ans were selling apples on the street, and people were singing "It's Only a Shanty in Old Shanty Town," and "Brother, Can
You Spare a Dime?" Unemployment had given America a new
leisure class, and even though they were without funds, they
were rich in resourcefulness. And that is the stuff of which
radio amateurs are made. Doggedly they scrounged, improvised, and built operating stations.

The January issue of QST carried a story about Navy
Day, the 27th of the previous October. This had been the sev-

enth annual observation of respect for hams by the United
States Navy, and was celebrated by significant nationwide relay communications. There were messages from the Secretary of the Navy, and from Hiram Percy Maxim, president of
ARRL. There were congratulations to the operators in each
Naval District showing the greatest proficiency in copying messages. A typical comment on this occasion (from W9COS) was,
"This is a most enjoyable work, and I am pleased to hear the
fist of our president [Maxim] at the HDQ's key at least once a
year. Truly, he sets a fine standard for us to follow, his key
work is admirable."

The July 1932 issue of QST gives an account of the meeting of the Board, "around a large oval table in the Hartford
Club." According to the report, they worked for two days,
morning and night, with time out for only a brief lunch and
dinner. The article explains that these board members "are
as different as amateurs can be, since ham radio is the meeting ground for every walk of life." It further emphasized that
these board members had traveled thousands of miles to merge
their views and judgments for the good of all. This attitude
has always been the keynote of ham activities.

During these difficult but momentous years, HPM continued as president of the ARRL, and writer for QST. In this
capacity he authored many beautifully-phrased and inspira-
tional editorials, as befitted the presiding genius of this technical field. However, there was still that other side to his
creative ability which remained a perfectly-kept secret until
after his death... those articles signed simply "T.O.M." (The
Old Man). These tirades usually started out with the admoni-
tion, "Say, son," and featured a caricature of an old fellow,
earphones lying near, and a tiger cat rubbing against his cheek.
T.O.M. was smoking a pipe, through his bushy whiskers.

The April 1930 publication of QST had carried one of
these articles, with the subhead, "In which our oldtime friend,
The Old Man lets loose some vital statistics about Radio Mavericks." It opened forcefully:

"Say, son, now listen... I've been gumming over this outside-the-bands business some more. It

seems that busting the transoceanic phone is becoming more and more fashionable. It's actually
got to the point where the phone companies have
logged so many amateurs on their phone channels
that they have plotted curves and can predict to
within a small percent the number of amateurs
there will be splashing around in the 14,000 and
the 6,990 channels at any time of the day or night
... Don't ask me how I know this, but I do just the
same. Hundreds of amateurs have been logged outside the amateur bands, and written down in the
black list, and when the lawyers say a rotten enough
case has been built up, the word will go out and
some people will have their tickets pulled.

"How is it going to look when K. B. or our President [actually Maxim himself] goes up against the
authorities, fighting our battles, and puts up a yarn
about how the amateur is so doggoned smart he can
build himself a radio set out of a few hairpins, to-
mato cans, and some worn-out dry cells, and in a
hellspitting hurricane send out radio messages and
get help, and some wisecracking congressman asks
how come, if he is so doggoned smart, why doesn't
he know how to stay in his own bands? It's going to
sound real musical, isn't it, when the Chief and
K. B. have to acknowledge that the amateurs have
busted the transoceanic phone service so many
times that a hundred or more amateur tickets had
to be pulled."

This was ARRL President Maxim speaking to the hams
in a language they couldn't fail to understand. It was a totally
different approach from that of the formal after-dinner speaker of the Hartford Club, and the Paris banquet room.

T. O. M. continued: "Now let me ask you a few
questions. Why is it there are so many of us over
the fence? What kind of ham is it that is found off
the reservation? What answer do they usually give
to the A. W. O. L. charges? What's the general at-

titude of these mavericks? No one can tell me it's
just plain cussedness. It isn't in the blood of most
of us hams to be cussed. The Wouff-Hong and the
Uggerumph and the Rettysnitch attended to this
years ago."

These three strange words which Maxim, in the guise of
T.O.M., uses here, refer to three mythical instruments of
torture that he "invented" for the disciplining of careless hams.
He made up fantastic stories about these weapons and had them
published in QST. The purpose was to make corrections in
such a manner that they would be remembered. The method
was effective. Every ham knows about the Wouff-Hong. A
crude model was made, and hung on the wall at ARRL headquarters. It is enshrined there to this day, richly framed and
mounted against red velvet, the focal point in their historical
museum.

T.O.M. continued with his tongue lashing:

"... There are a lot of us suffering from extreme
youth and the lack of the wherewithal to go and buy,
or the opportunity to otherwise acquire, a precision
heterodyne wave-meter. Those mavericks who
stray off the reservation simply don't know where
the reservation lines are. All right, then it's up
to us older heads to mark the channels somehow,
so that it's easier to stay in them. How are we going to mark them? We amateurs can't expect pointed buoys or flashing lights, and the depth in fath-
oms. Any old kind of stake or bush indicating the
limits of the channel is good enough for us. Now
why not get up some kind of marking for the limits
of the various bands that we use? Let ARRL maintain these 'radio lighthouses.' Then the young
squirts, who have to get from Dad the underwriting
of funds for everything they buy, and who are too
busy calibrating their receivers, can tell at once
if they are over the fence or not."

There is even less of the polished gentleman in his closing paragraph:

"And then, by gorm, every maverick that is caught off the range will get branded by the Radio Supervisor, and this brand will begin with a big R, and this big R will mean rotten. Wouldn't it be elevating to use a call beginning with an R?"

These wonderful explosions were eagerly read by the hams as soon as they flipped open the pages of a new issue. And the ideas they conveyed are remembered, even to this day, in amateur circles everywhere.

In contrast to the secretly-penned T.O.M. articles are the scholarly, formal editorials presented in QST under the signature of Hiram Percy Maxim. An example of the president's writing is this selection from QST of January, 1932:

"We radio amateurs are the champion 'Just Supposers' of all time. Our entire structure has been built up by just supposing. By just supposing, we broke into "wireless" immediately when Marconi showed us that there was such a thing. By just supposing, we dug up what had to be done to make the useless two-hundred meter wave useful. By just supposing, we discovered that a strong brotherly spirit could be created, out of which we built ARRL.

"By just supposing, we found a way to use one hundred meters and work two-way across the Atlantic with certainty. By just supposing, we found out that there was a way to use forty meters, and then ten meters, and then five meters. By just supposing, we were led carefully to pick over a lot of different men and finally select those whom it would pay to train.

"My own long years of experience as your president are an example of just supposing.

"... the standing of amateur radio and of our ARRL before the government and the public at large is higher than ever before in history. The standing of amateur radio in foreign countries, heretofore all but hopeless, has begun to show the most encouraging signs in all history.

"Now just suppose this steady advance is maintained; that all nations of the earth continue to get better and better acquainted with each other; that fear of uprisings and revolutions grows less and nations gradually relax their absolute control' of communications and permit their citizens to communicate freely with each other as we do in America. Isn't it possible that what we now call amateur radio may really become world-wide? That children in school will be taught the telegraph the way they are taught to read and write? That the day may come when almost everybody in the U.S.A. is independently hooked up by citizens radio the way almost everybody in the U.S.A. is independently connected up by good roads and automobiles?"

*   *   *

By 1932, amateur radio had been well established in the world, and its presiding genius, besides growing a bit older, had developed a new interest, namely the infinite possibilities of the cosmos. Accordingly, he became a little less active among the radio amateurs. He made a globe of Mars, and began work on a book, **Life's Place in the Cosmos.** Occasionally he would flash out of his semi-retirement, onto the pages of QST. Such a circumstance occurred in the February issue, perpetrated by T.O.M.:

"Say, you modern young Squirts... in bygone years, when amateur radio was young and in the making, it was unpopular to have a rotten fist, send with your feet, or send too many C.Q.'s. It was unhealthful, very, to be outside the regular amateur bands. It was Sure Death to fake a call or send any kind of a false signal.

"There were three gadgets that were devised by the amateurs of those early days to keep young Squirts constantly reminded of these important don'ts of amateur radio. A Squirt who used too much of what we used to call 'Lake Erie Swing,' or send with a slobbery fist, or clutter up the air with too many C.Q.'s, or garbled his call letters so they had to be guessed at, was called upon by a committee, the chairman of which was a big brute with a positive manner, and who exhibited and explained the workings of an instrument known as an Uggerumph.

"A Squirt who was a band-jumper, or who failed to maintain an intimate acquaintance with a reliable wave-meter, was politely knocked on the head with a baseball bat, dragged out into the nearest sandlot, and subjected to a thing called a Rettysnitch. The blood of each victim was allowed to dry on this tool. She's all caked up on the business end this minute. The next time you meet your old friend Fred Schnell ask him how many Squirts he butchered the first year or so after the war, with this Retty snitch. He nearly wore it out. He was Traffic Manager in those days."

There is little here to remind one of the polished pleadings of the ARRL president before the Senate Committee in Washington. But Maxim fitted his words to the need. He continued his tirade against those thoughtless operators who clouded the good names of the countless innocent:

"Last but by no means least, a Squirt who even thought of using a false call, let alone actually using one, or using profane language on the air, or who willfully broke up other legitimate amateur traffic, was taken for a certain kind of a ride in which an instrument of torture known as a Wouff-Hong figured very prominently. No Young Squirt ever returned from one of those 'rides.' The Wouff-Hong was carefully wiped off after the affair, and the rag used to do the wiping was forwarded to me for filing away. There are twenty-five bales of these bloody rags in a certain warehouse at the

present moment, and recently I have made arrangements to hire some additional space.

"Now I don't want to be unpleasant or to threat-
en anybody, but by the Great Horn Spoon, you Modern Young Squirts, you are sure riding for a fall if
you keep on the way some of you are going. The
other night I counted one of you sending forty-one
CQ's and then, to add insult to injury, you ended
up with just two miserably-sent slobbery signs...

"Now I ask you, why forty-one CQ's if two un-
readable calls? What sort of a think-tank do you
carry around that led you to sign only twice and
rotten at that, if you thought forty-one CQ's were
called for? But I must not permit myself to dwell
upon your offense for fear this paper will get afire.
Just let me say that I've got my eagle eye upon you
and any moment you may expect to receive a call
from my committee, the leader of which will take
such steps as may be necessary to convey to you a
clear and distinct idea of just what the Uggerumph
is, and what it can do, how it is operated, and how
it feels."

This sort of clever plain talk got the needed results far
better than preaching would have, and kept many a thoughtless
ham from losing his license, or getting into serious trouble.
This self-regulation and inside-discipline was the secret of
keeping the short waves clean.

The Old Man, like every amateur today, enjoyed his hours
of informal "rag-chewing," and was every bit as human in his
reactions as anyone when some inconsiderate operator cut him
off:

"Another one of you washed out a perfectly good
'rag-chew' for me and ditched several words from
an important QST from W1MK. And you, you rubber-
head who did this, you are going to be hounded down
and I'll see that you are made to eat every V in the
dictionary. In the old days we would have had you
boiled in transformer oil. I reckon you are too

young to know what transformer oil is. Well, Son, it's hot, when it's boiling.

"Another one of you thinks it funny to send with your feet instead of your hand. It cannot be done, Son, and it isn't funny. It's rotten amateur radio. Nobody can read your stuff, and what's the use of sending it if nobody can receive it? If your stuff were to be taken down on a tape receiver it would bust the machine wide open and hurt somebody. There are only two characters in the telegraphic code. One is a dot and the other is a dash. You seem to think it fashionable to cross-breed the two. It can't be done without getting into trouble, and you are in trouble because some dark night you are going to be yanked from your warm little shack and a certain committee will heat up a pair of irons and several real distinct dots and dashes will be branded upon that part of your anatomy which will remind you of dots and dashes every time you sit down, for the rest of your life."

This was in contrast to the dignity of legislative bills, and academic lecture platforms, but the Chief was probably never more effective:

"... I have already sent to Hartford for the Wouff-Hong to be forwarded here by airmail. You enemies of amateur radio will soon be hearing the old pre-war torture chanty... a horrible example will be made of you for the benefit of future Young Squirts who are tempted to stray from the straight and narrow."

These tongue-in-cheek harsh words were penned in the hope that they would prevent some thoughtless ham from losing his license, and then having that fact published for all to see, as was done in those days.

"I have not yet decided what the torture shall be for those miserable whelps who splash around in the ether and get afoul of the transatlantic phone bands. It's a serious offense. It gets all the rest of us in dutch with the government and A. T. and T. This matter is up with the Torture Committee of the ARRL, and as soon as something sufficiently bloodcurdling can be devised, we shall rig up the apparatus and bring you whops in.

"I am in favor of stringing all of you up by the thumbs in good old seafaring style, while somebody with a deep voice reads one of Handy's papers on Dynatrons and Methods of Monitoring Frequency, after which all the worn-out and inaccurate wave-meters in your Division will be crammed down your throats.

"The idea is to make you 'frequency conscious. I reckon it will make you conscious all right, especially about the time you feel some of those crook-ed condenser plates going down.

"One of the suggestions the Torture Committee is working on is to have Warner (ARRL Secretary) suggest to the forthcoming Madrid Conference the advisability of compelling all amateurs to eat at least one dynatron before he learns the code."

This is high humor, but it also made good sense to the hams. They talked over what The Old Man said, and they respected it.

"It has been a long time since this old barnacle has opened his trap, but after things reached the pass where one of you Young Squirts pinched the Radio Supervisor's call, I had to speak. I am here to see that amateur radio retains its traditions, built in bygone years when amateurs were loyal, law-respecting and upstanding he-Americans, who were proud of their fists both on the key and in the law breakers' faces, proud of their procedure, proud of their frequency, proud of their delivered-messages percentages, and proud of the splendid standing of amateur radio. The Uggerumph, the Rettysnitch and the Wouff-Hong have served in the past, well and terribly. They can serve again."

"You present generation of Young Squirts are
too young to know the past... but believe me, there
was a past."

These words make it apparent why Hiram Percy Maxim
is called the Father of Amateur Radio.

It was about this time that Maxim began the writing of
Horseless Carriage Days, and A Genius in the Family. These
were humor-tinged autobiographical accounts. The latter was
eventually made into a movie called "So Goes My Love," star-
ring Myrna Loy and Don Ameche. Bobby Driscoll played the
part of young Hiram Percy Maxim. The family was disappointed in the film because of its departure from fact.

HPM was also involved in writing a regular column for
the King Syndicate. It was a series of pseudo-scientific articles written "for the street car conductor, and the policeman
on the beat," as the author put it. The subjects included horseless carriages, noise and space. There were comments on
morality and immortality:

"The intriguing thing about life and living matter is that it is to the last millionth of a percent
nothing but inorganic or mineral matter. It seems
as though it were the bricks which are immortal,
and the structure or combination of bricks which is
mortal."

Many times during his adult life, Maxim thought about
that childhood day at the Philadelphia Centennial, when he had
gazed at the meteorite. He was entranced with the idea that it
was millions of years old, and that it had come to this earth
from the unknown regions of outer space.

More and more, as he grew older, he contemplated the
significance of the cosmos. Eventually he embodied some of
these thoughts in an article, which was published in Scientific
American, in April of 1932. In it he stated:

"One of the questions that occasionally comes
to all of us is... what is this great scheme of nature all about... what could have been the beginning of this stupendous mechanism... where is it
going to end? Where does the phenomenon of life

fit into the great picture?"

He goes on to explain that these are questions that mankind has been asking himself for centuries. The answer, he felt, might eventually be found as mankind unveiled for himself the secrets of SPACE. The significant fact to him was that space seemed to be an inorganic universe, for we have been unable, in all our searchings, to find one shred of positive evidence of the organic there. So many conditions in space seemed just the opposite of those prevailing on earth, which is always teeming with things organic, and which are able to reproduce their kind. Further, so far as we have been able to discover we have a unique intelligence here on earth. But the intriguing aspect of this intelligence to Maxim was that it is, comparatively speaking, so newly-developed. Maxim believed that man's ability to build airplanes, railroads, telescopes and the like, and to understand many of the complex laws of the universe, was just the beginning of much greater things. He went even further, saying, "This earth may not be the single abode of intelligence in the domain of space, among the billions of celestial bodies." He points out that according to the accumulated wisdom from Ptolemy to Gallileo, and thence to Sir James Jeans, that we humans seem terribly alone and adrift in a transcendentally vast ocean of time, space and energy. "And the evidence," he notes, "is that we have been thus for a very long time... for over millions of centuries. Where we are going, and why, fills us with immense wonder. Our Parent Sun is quite an average star, and more staggering thought, there are thirty thousand million suns, and hundreds of thousands of galaxies. The fact that we are the inhabitants of only one of nine planets of one single sun, leads us to wonder if we are quite as important as we sometimes think."

When questioned about the possibility that there might be life on other planets, Maxim was judiciously thoughtful. If life existed on Mercury it would have to be between a killing heat and a killing cold, and so far there is no evidence to support the possibility. Venus, Maxim believed, might have been sending us radio signals for years, which we are as yet too backward to detect. "But only in a very narrow zone," could the kind of life exist there that we can imagine. Ever the radio man, Maxim stated, "Radio waves represent our first tool with which it may prove possible to carry a signal across the great reaches of astronomical space."

"If life does exist somewhere else," he continued in his 1932 magazine article, "and it is reasonable to expect that it does, then some day someone is likely to encounter, by means of radio, an extra-terrestrial intelligence. What a sublimely dramatic moment it will be for those concerned when this first interstellar contact is made! Will it not dwarf every other dramatic incident that ever happened upon this earth?"

Maxim pondered too about the language barrier. "Think of asking 'it' where 'it' resides... on what celestial body 'it' is located. And what if 'it' were to turn out to be an intelligent insect sort of thing, or an intellectual vegetable. It might easily be." All of this sounds like the science fiction which is fast becoming reality today.

Maxim points out that while radio will be our first means of contact, it has certain definite limitations. For instance, it would take forty thousand years to cross the space between Earth and one of the stars of the Hercules Cluster... and another forty thousand years for an answer to come back!

In 1932 Maxim proclaimed, "We are on the threshold of things. If we go on advancing in accumulated knowledge as we have advanced since the time of Columbus, our future condition will be impossible to imagine."

Then, as if he glimpsed our atom age, with all of its horrible possibilities, he added, "If no accident happens, and if human beings go on developing and advancing in knowledge even faster and faster, what is to be the influence of intelligence on the cosmos? Is it possible that intelligence is the big aim and object of creation? As we gaze into the sky on a clear night we are filled with a great feeling of reverence, for we realize that we are looking at the Great Machine in operation."

Maxim became especially interested in the planet Mars, and its possible relationship with Earth. He made a globe of the planet, complete with physical features as he ascertained them through telescopes. His chief interest in the Percival Lowell Observatory in Flagstaff, Arizona, was its reputation as a recognized authority on Mars. This institution had been established in 1894 by a non-resident professor of astronomy at MIT, Maxim's old school. In 1904 the Observatory had been awarded a medal by the Astronomical Society for their research on Mars.

By 1933, Hiram Percy Maxim's book, **Life's Place in the Cosmos,** was taking form. He wrote:

"Of all the dramatic situations that imagination can picture, the most dramatic, it seems to me, is presented by the creature man, perched upon his little speck of cosmic dust, peering out upon a hostile cosmos, fully aware of his own physical insignificance... yet boldly digging out of an omnipotent and recalcitrant Nature one after another of her treasured secrets, ruthlessly disclosing truth after truth and boldly exploding century-old superstition after superstition."

That was the Maxim mind, probing, eager. There were a number of good reasons, to him, why we human beings sometimes marveled at our own importance, and at our own accomplishments. He mentions a few of them: the construction of the Empire State Building, with a height ten times its width at the base; the building and operation of a ship which is a fifth of a mile long, yet in no danger of breaking in two; the fact that we can separate continents and join oceans; and the ability we have of weighing the sun and the stars, and determining their dimensions and temperatures. Truly, these are great accomplishments, but Maxim had a way of reducing even these to their proper proportions... "Step outdoors and look skyward on a clear moonless night." This, Maxim explained, was a refreshing diversion from "hackneyed thoughts of money, work, and worldly matters." The question uppermost in his mind in this regard was "what the Great Plan of Nature may have in store for us who possess what appears to be a monopoly of intelligence."

Maxim pictures prehistoric man behaving, in this regard,

just as we do today, and have done, all down the years...gazing and wondering at the spectacle... "before our eyes spreads the black vault of which we know so little but which contains the key to the mystery. It is spangled with myriads of twinkling points of light."

Vast and majestic as is the beauty of this, there is an even more awe-inspiring angle, that which concerns the nebulae that are a million light years away. Maxim explains: "A photograph made tonight shows what they were a million years ago. There is no possible way of knowing what things are like in the cosmos NOW. Cosmically speaking, there is no such thing as NOW. NOW HERE means something. It means the present time on Earth. But if we look across the cosmos NOW at a nebula a million light years away, we see the nebula as it was a million years ago. NOW HERE on Earth is a million years different from NOW on Nebula M31."

Maxim links this with Einstein's "time means nothing and space means nothing," explaining that accordingly, the only thing that has real meaning is WHEN-WHERE, or more commonly known, 'space time.' It is an idea such as this which provides the plot for so many television scripts, and it is the field in which fact is so rapidly overtaking fiction.

Of the whole spectacle, Maxim states in his book, "Our insignificance seems to be magnificent." With a knowledge based on years of study and observation, he discusses the planets, the stars, the Milky Way, and the galaxies. He writes:

"If Einstein is right, and we had a powerful enough telescope, we should be able to look 'around' the cosmos and see our Milky Way galaxy some five hundred million light years distant. We would be looking the wrong way around. We would be seeing our galaxy as it looked five hundred million years ago."

And, he adds:

"The astronomer Jean says that 'stranger things than this have come true in astronomy.

"We are able to perceive only three-dimensional space... but there are four dimensions. The

fourth is space-time. The first three are physical,
> the fourth is a purely intellectual concept."

Thus it follows, Maxim explains, that time is not a reality. What we call time is really only one event followed by
another one. In relation to the cosmos, we cannot limit our-
selves to a space that is physical, because the moment we do
that, we have erected a boundary.

Having outlined in as simple terms as possible the physical aspects of the universe, Maxim turns to a contemplation
of MAN himself:

> "No creatures have ever existed on Earth that
> equaled man in intelligence. We represent Nature's
> best. And so the best that Nature has been able to
> do in something like three million years is what we
> see in the streets today... wherefore, we pause and
> wonder anew, what is the status of Intelligence in
> the Great Cosmic Scheme?"

Maxim goes on to expand the picture:

> "... And you and I, standing out here under the
> great vault, sub-microscopic organisms, crawling
> about upon a microscopic speck of cosmic dust, a
> minute fragment of one grain of all the grains of
> sand; and yet withal possessing a 'something' which
> enables us to detach ourselves and abstractly view,
> analyze, and appreciate the affair, dispassionate-
> ly, judicially, critically... Notwithstanding we
> have no command over our own existence and may
> be swept to extinction in the next few minutes. But
> WHILE WE DO EXIST, this 'something' makes us
> independent, unafraid, and superior."

Thus in this book we have Maxim the astronomer becoming Maxim the philosopher, enjoying the picture of a 'speck of
dust' daring to stand up and criticize the universe.

From this position it was only a hop and a skip to HPM's
favorite subject of radio:

> "And now let us have a look at the question of
> communication, for it would be tragic indeed were
> we to know that fellow intelligent creatures were in

existence but we might never communicate with them.... communication across outer space can be accomplished only by means of the agency we call 'radio.' At least we can conceive of no other possible agency."

At this point in the book, Maxim introduces the history of radio up until that time, pointing out that we have no advanced very far in it because of the short span of time since we first knew much about it, in 1912. He recounts the story of how the amateurs took their assigned territory of two hundred meters and down, the then-useless meters, and explored and conquered them. His theory was that this process could be continued, and that below twenty meters there is theoretically an infinite number of wave lengths before the point of zero is reached. Visible light, Maxim explains, begins at red light, with a wave length of .00008 centimeters, so there is a great interval between five meters and the beginning of visible light. He thought it entirely possible that somewhere in this interval there could be signals sent out by an extra-terrestrial intelligence. He stated that there were all manner of electromagnetic oscillations existing, but that our techniques in detecting them are probably too clumsy to judge whether they have the characteristics of being originated by intelligence. He dreamed of the day when a radio operator here on Earth would be able to sweep through the wave length and detect strange signals with some sort of system about them. He writes: "Then we shall know that some intelligent creature somewhere not of this Earth is calling."

The book then recounts the story of a strange offer which originated in France. In 1893, the Pierre Gusmann prize of 10,000 francs was established, to be awarded to the first person to establish contact with an inhabitant of any extraterrestrial body. This unusual legacy was accepted by the University of Paris, and they still hold it. The exact conditions are "to send some sort of a code to one of the stars and receive some answer back from there."

At the time that this prize originated, Hertz had just announced his Hertzian wave researches, and the press was enthusiastic about the possible aspect of distant communications.

Maxim, in commenting about the Paris prize offer, stated: "Unfortunately, the world has suffered some serious disturbances since then, and the prize which originally was twenty-thousand dollars in American money now amounts to but five thousand dollars, owing to man's defective political and economic civilization."

Then, even allowing for all the wonderful advancements which man might make in the field of radio operation, Maxim comes head-on into a hard fact:

> "But we must consider the sad limitations that cosmic distances place upon communication. The velocity of electromagnetic oscillations or light is 186,000 miles per second. Nothing can ever travel faster... when we come to cosmic distances, the time for light to cover them can become years instead of fractions of a second,"

as here on earth. In point of fact, it would be possible, under favorable circumstances, to communicate with Mars in half a minute, and that would present no serious time problem.

> "But now comes the sad part of the story. To find the next celestial body that promises any hope we must leave our Solar System. We must move out to the nearest fixed star. It is Proxima Centauri, distant some four and a quarter light years."

Maxim points out that this would mean a four and a half year lag in communications, and suggests that this would cramp the style of those desiring to gossip. He then emphasizes that if we are selfish, we will be content to just drop the matter there. But he himself was not ready to give up. "We must see this thing through," he urged. "We should go on contriving, delving, improving, helping and imagining." It was this quality of inquring persistence that made HPM a successful scientist and inventor, witness the fact that he discovered the principle of the Silencer by watching water swirl down the bathtub drain.

In the final pages of **Life's Place in the Cosmos,** Maxim takes a look at the future:

"As a result of improvements made in transportation and communication the races of the entire
world are interbreeding for the first time. In a few
thousand years complete infusion will have occurred, and there will be an absence of sharply
defined races.... whether society crumbles or not,
the animal man will probably survive. He may have
to go fairly far back and start over again, but he
can do it if necessary and profit by the experience
of the race."
Brought out also in the book is Maxim's theory that inventions develop in relation to man's needs, and that one invention begets another. Thus the bicycle served until our way
of life demanded a car. That mode of transportation satisfied
until mankind needed something still better, and devised the
airplane. Maxim speculated on what would come next. He was
ready to accept the idea of atomic power, and of the missile,
and the man-made satellite. He urged, "Let us do our humble
bit in encouraging the collection... of fact."
All of these ideas Maxim wrote into his book in 1933
when we were being affected by a bank holiday, the NRA, the
CCC, and a host of other alphabetical innovations. But we
were singing a lot of good songs too, like Hoagy Carmichael's
"Lazy Bones," and Jerome Kern's "Smoke Gets in Your Eyes.'
That was the year Irving Berlin gave us "Easter Parade." It
was the year Jimmy Durante introduced "Inka Dinka Doo." It
was the era of strong rhythms like "Carioca," "I Like Mountain Music," "Little Grass Shack," and "Who's Afraid of the
Big Bad Wolf." Coincidental with Maxim's cosmic theme,
1933 was also the year that brought us "In the Valley of the
Moon," "My Moonlight Madonna," "Orchids in the Moonlight,"
and "It's Only a Paper Moon."
In this year of economic shambles, with business down
to sixty percent of normal, with exports at their lowest ebb in
thirty years, with one thousand four hundred bank failures over
the country, and unemployment estimates of thirteen million,
the one bright spot in the nation was radio.
For the amateurs it was boom time, with over eleven

thousand new stations licensed in this one year. There were several good reasons for this paradox. Men and boys were unemployed, or able to work only part time, and had new leisures to fill. Secondly, with broadcasting and its associated fields clobbered by the depression, prices of radio equipment went down, through cut-throat competition. Also, research was improving the quality. The broadcast-receiver builders turned to short wave as a newer and more fascinating field. At this time too, came the opening of the ultra-high frequency bands for amateur use. Following this, many services turned to the short waves to help solve their problems. Among them were police departments, short-distance radio-telephone facilities, forest fire-fighting units, construction projects, and overwater telephone circuits. Added to this was the whole new era of television, which was just emerging from fantasy to reality.

On March 29, 1933, Herbert Hoover wrote a letter to Maxim. In it he said, "I am delighted that the League is making such progress."

In April of 1933, newspapers and magazines carried items and reviews of HPM's new book. One of these featured a photograph, with the caption "Hiram Percy Maxim, inventor, and author of Life's Place in the Cosmos (Appleton, 2.50) is shown working on his globe of Mars."

Another publication carried a cartoon of HPM talking into a microphone, with a helmeted Martian answering, "Howdy," The caption read, "Maxim puts in a long distance call."

Maxim's employee, Sundkvist, at the silencer plant, was not so complimentary. He termed the book "almost irreligious in spots." To which HPM replied, "But we have a Supreme Being."

On August 5, 1933, Maxim addressed the ARRL convention in Chicago:

"History is being made here tonight. This is probably the most important gathering of radio amateurs the world has seen... The Martians had a head start on us of many millions of years..." [he went on to explain why he thought so]. "Communication with Mars will be the most dramatic that

has ever occurred on Earth... here's hoping he
will be an amateur, and a member of good old
ARRL."

The year 1933 also saw the Madrid Convention on Amateur Radio, with provisions later ratified by our Senate. In
October there was a complete revision of the regulations of the
Federal Radio Communications. While many details were
changed, actual operating procedures were left practically the
same. Three forms of licenses were established. Mobile operations and informal portable set-ups were permitted. As
Clinton B. DeSoto says of this period: "Expansion in knowledge
and technique took place both vertically and horizontally at
tremendous rates. There was continual pioneering in new
fields."

Some of these new things led to new problems. People
with radio receivers in their automobiles began picking up police broadcasts, and rushed to the scene of the crime or disaster. Too often they arrived ahead of the proper authorities,
to the danger and confusion of all concerned. This brought on
legislation at a municipal level, and it was at this point that
the ARRL stepped in to see that the licensed ham mobile equipment was a matter for federal regulation alone.

In spite of the downward economic trend, 1933 was an
important year for expeditions and exploration. Two outfits
were in Greenland; two others were in the Antarctic. Commander Frank Hawkes, a dedicated amateur, used radio communication on a non-stop flight from New York to Regina, Sas-
katchewan. A party in Alaska used amateur radio in mountain
climbing, and another group in South America used radio to
explore some placer gold diggings along the eastern slopes of
the Andes. U.S. Soil Conservation crews with the Colorado
River Expedition carried amateur radio equipment. A group
from the American Museum of Natural History working in the
interior of Brazil used amateur radio in the transfer of supplies and specimens.

Terrible emergencies occurring in 1933 saw the use of
amateur radio to help thousands. Among these was the Southern California earthquake in March. Within ten minutes after

the disaster struck, ham operators were on the air with salvaged tubes and parts, and emergency power-supply facilities.
They helped thus during the four hours it took before any of
the local broadcasting companies could get back on the air, and
during the six hours that elapsed before wire line service
could be restored. August of 1933 saw floods in Virginia,
Maryland and Delaware, with amateur radio the sole communication for workers, officials, and the press. That fall there
were hurricanes in Florida and in Texas, with amateur radio
again bringing help to the marooned and stricken areas.

The winter of 1933 found amateur radio operator Claude
DeVinna in remote Alaskan country with an MGM motion picture crew which was filming "Esquimo." His station was located in a little hut offshore from his supply steamer, which
was frozen in. While DeVinna was communicating via ham
radio one cold night with another ham in New Zealand, he
was overcome by carbon monoxide fumes from the hut's little
gasoline heater. The New Zealander, when the Alaskan signals faltered, sensed that something had gone wrong, and,
still by radio, dispatched a relief party from Teller, Alaska,
just in time to save the unconscious man from death.

Thus 1933 was the year for amateur radio to be publicly
appreciated and respected for what it is, a universal, unsel-
fish service, as its founder Maxim always emphasized.

June 19, 1934 saw the passage of the Communications
Act. It created the Federal Communications Commission, re-
placing the Federal Radio Commission. The structure and
definitions, however, remained practically unchanged. During
the four years that had elapsed since Maxim's speech before
the Washington Senate Committee, more great things had happened to amateur radio. There were new inventions, new
techniques, and thousands of new operators, all of this in spite
of the Great Depression.

There were other bright spots in the country too. Little
Shirley Temple, with her deep dimples and bobbing curls had
enshrined herself in the hearts of the nation. People were singing some wonderful new songs in spite of their troubles...
beautiful melodies like "Blue Moon" and "The Isle of Capri"...

haunting tunes such as "Solitude" and "Tumbling Tumbleweeds" ..... and the rousing rhythms of "Anything Goes," and "Blow, Gabriel, Blow."

This was a generation of polarities... crushed and hopeful, despairing and exuberant. It was a time when people had a great deal of forced leisure, and when a good radio outfit could be had for less than fifty dollars, or improvised from salvaged materials. The applications for licenses mounted daily, practically the only growing thing in a nation of drouth.

Personally, the Depression had hit Maxim hard. He had a fearful time saving the Silencer Company. His wife was not well, adding to the anxiety and the expense. Since radio would have meant an expenditure, Maxim's activities in it were curtailed at this time, but he carred on as President of the ARRL, and as head of QST.

It was during this period that Maxim was invited to Corning, New York, to witness the pouring of the 200 inch mirror lens for the then-largest telescope in the world, at Mt. Palomar Observatory. He recorded his thoughts on this occasion: "You and I and yours and mine seem to be inhabitants of a little speck of cosmic dust, which is a satellite of a slightly larger speck of incandescent matter, which is in turn one of an unknown number of galaxies.... our insignificance seems to be magnificent."

During 1934 the country seemed to have more than its share of physical disasters, and in all of them amateur radio came to the immediate rescue with medical help, food and supplies. There was storm and flood in Washington, Idaho, Oregon, North Carolina, Arkansas, Tennessee and Mississippi. There was a plane search in the Adirondacks. There was wind and sleet in British Columbia, Maryland, Delaware, Virginia, Wisconsin and Minnesota. With all other communications cut off from Superior and Duluth, amateur radio instantly took over the job, and was used by the Northwestern Bell Telegraph Company, Western Union, Postal Telegraph, a power company, and even by a brokerage house.

Late in 1934 the first amateur "round the world flight" was accomplished by Dr. Light and Robert Wilson of Yale, who logged thirty thousand miles in their cabin plane out of New York. It was another milestone in communications history.

Through all of these diverse experiences, amateur radio earned commendation everywhere, from grateful individuals, and from the Army, the Navy, and the Red Cross. The amateur operator was praised as a free agent, acting at his own discretion, serving voluntarily and without compensation, save the thrill of achievement, and the satisfaction of a job well done. Maxim's dream for his organization had been fulfilled.

This was the period when Maxim had so much fun being a grandfather, or Fäd, as the children called him, and Percy had before that. He was very successful in the role, and always had time to listen to a child's stories. On one occasion Percy's little son Jack and his younger sister told a wildly exaggerated dinosaur story to HPM. He recounted right back with a very serious face, until Jack's conscience began to hurt, and he said, "Fäd, that wasn't really a true story."

Maxim hid his amusement and answered confidentially, "I'm glad you told me it wasn't true."

Young Jack passed many happy hours in the Maxim 'conversary.' To him it always seemed dimly lit, and since his grandmother always sat with a book open on her lap, he came to believe that she was the only person he knew who could read in the dark.

Sometimes HPM took his little grandson with him to the Silencer Plant. On one such occasion he put him in charge of one of the men to watch the welding, so that he would be kept pleasantly and profitably occupied.

*   *   *

In 1935 Americans had forty-one new hit songs to raise their spirits, including Cole Porter's "Begin the Beguine," and Gershwin's "I Got Plenty of Nuttin'." People were learning to live with the Depression.

Mrs. Maxim, true to her spirit of humanitarian dedication, was serving her third term on the Public Welfare Committee.

On September 7, 1935, the Hartford daily carried an article by Maxim on a subject most unusual for him:

"Religion to me means the enobling influence that affects the human mind as it contemplates the sublime law, order and grandeur of creation. The more that is revealed of the great plan upon which creation is erected, the deeper the awe and reverence that we feel for the Architect who designed it."

"As man grows in his understanding of what confronts him on his little earth, and in the heavens, the more he finds there is to know. He has but begun his understanding... as the ages unfold, and his understanding increases, so will grow his religious reverence."

Maxim liked to write fanciful stories about the possibilities of the future. One dated December 18, 1935, is titled, "Dodging the Grim Reaper." It is in his own handwriting, on Mayflower Hotel stationery, Washington D. C......:

"How would you like to go to sleep and wake up a century hence? If you repeated these century naps until you were 70 to 75, you could have lived in twenty different centuries, and could spend your declining years around 3950. You would be in the position of a man who had lived in Julius Caesar's time, a few years before the birth of Christ."

He goes on to imply that this sort of suspended animation could help us await cures for currently-fatal diseases.

December, 1935, marked the first improved color movies that Maxim took. The reel includes shots of his grandchildren engaged in a lively snowballing, to the excitement of their two dogs. There are also some interior scenes featuring a Christmas tree and wreath. Maxim had developed a device so that he could appear in pictures he took himself. It was used here so that he could be filmed, half squatting on the floor watching the grandchildren play with their toys. He opens his gift from his little granddaughter, and kisses her. It is a family observation that HPM was happily demonstrative, whereas his wife was reserved, and almost shy. In this Christmas scene, Josephine is seated in a wicker chair, appearing frail and tired. The camera records the opening of her gift, a family photograph.

This was to be their last Christmas. Within less than two months, both of them would be dead.

Early in 1936, HPM was presented with a new rig, but he never got a chance to use it. This equipment still has all but four of the original tubes, and is in the ARRL museum at QST headquarters in Newington, Connecticut.

When the New Year Babe of 1936 arrived, radio was still one of the brightest spots on the economic horizon. Things were looking somewhat better at home, but there were rumblings from abroad. We had experienced the WPA, Huey Long, the Dionne Quintuplets, and the Social Security Act. We were familiar with Fireside Chats, Oakies and crooners. We had forgotten some of our troubles and had begun to hope. Even the songs were optimistic. We chorused "Isn't This a Lovely Day," and "You Are My Lucky Star."

It was in 1936 that Clinton B. DeSoto brought out his history of radio entitled **Two Hundred Meters and Down**, and in it he presents some prophetic ideas, including the use of amateur radio to communicate to Earth from a space ship, and the development of amateur television. He emphasizes his point by stating, "If it comes, when it comes, you will owe it in large extent to amateur radio... just as you now owe in large measure to amateur radio your broadcasting that entertains, your police radio that protects, and the... television that tickles your speculative fancy."

DeSoto presented a neat summary of ham progress:

"Nationally and internationally, from all standpoints, technical, fraternal and legislative, amateur radio occupies an enviable position in the year 1936. With a larger proportion of frequency assignments than any equivalent service, and prospects of securing more, a place wherein it can long live is assured. Recognized and perpetuated by national law and international treaty, backed by a wealth of precedent, stronger politically and technically than ever before, protected and safeguarded by uniquely successful national and international organizations, the right to existence seems certain."

DeSoto decided to dedicate his book to Maxim because he had organized amateur radio, nourished it, fought for it, and disciplined it.

In January of 1936, Maxim and his wife planned a vacation trip to the Southwest. One of his objectives was a visit to the Percival Lowell Observatory in Flagstaff, Arizona, where he had been invited to make some planetary observations. In order to be free for the trip, Maxim had put in a heavy stint of writing for the King Syndicate, to complete a series of articles. He was on the point of exhaustion when he wrote a letter to Roy Winton, managing editor for his column:

"I leave February 8th for a month's vacation. Be sure to send me anything upon which I must act, so that all will be in order before I leave."

There was heavy fog in the east, and blizzards in the Midwest when the Maxims left for Arizona. HPM, often a victim of throat and chest disorders, became ill, and was treated by a physician between trains in Kansas City. His condition worsened, and he was taken off at La Junta, Colorado, and put in a hospital. He was delirious, and talked almost constantly of the early car days, and of pioneer radio, as if he were still struggling with their problems. The family was summoned from Connecticut. The hams set up a relay from La Junta to Hartford, via Denver, and issued bulletins on HPM's condition. There were forty-eight hours of encouraging progress and hope, but then on February 17th, at the age of sixty-six, without having regained consciousness, Hiram Percy Maxim died.

The newspaper headlines of the day described the world he left: "Lawyer Grills Hauptman," "20,000 Ethiopians Slain in Fight in Northern Front," "Search for Cancer Cure is Simplified, View of Experts," "Smashing Victory for Rightists Claimed in Spain," "Townsend Plan Argued," "George Washington Legend Debunked," and "Country Awaiting Decision on TVA."

In the world of sports, Sonja Henie of Norway was in Olympic ice glory.

On the comic page, Little Orphan Annie was running away again, saying, "C'mon, Sandy, let's keep movin' along."

The entertainment pages listed "The Big Broadcast of 1936," with stars Jack Oakie, George Burns, Gracie Allen, Charlie Ruggles and Bing Crosby.

The fashion section proclaimed a trend toward dark sheers, novel prints, and slim gored skirts. For little girls, Shilrey Temple dresses were high style. Zipper overshoes were a new item, and cost 98¢.

The grocery ads offered coffee at 2 pounds for 55¢, butter 2 pounds for 59¢, bananas 3 pounds for 21¢, and oranges for 1¢ each. Flour was $1.39 for forty-nine pounds, and seven tall cans of milk were 49¢.

A popular song of that day was ironically symbolic, "Empty Saddles in the Old Corral."

The Maxim family was consoled by the fact that HPM had died in full flower, without having had to slow down, which he would have despised. They found inspiration in a prayer card which was in his pocket at the time of his death. It was titled "A Blessing," and provides an answer for those who had sometimes commented on his lack of church affiliation (although he gave handsomely):

"Almighty God, that which compels the reverence of all thinking minds, who has established the never-failing rule of law and order which prevails in all things, both earthly and celestial, we pause at this time to bow in religious retrospect. As we behold the wisdom, the justice and the scope of the great Scheme of Nature, we are made better men and women, and are led to endeavor to be more wise and more just and more orderly in our daily lives, and in our relations with those among us. Amen."

It was the Silencer firm's secretary, Cecil Powell, who

broke the news of Maxim's death to associate Sundkvist. She said simply, using her familiar name for him, "Papa died."

Several days later, at Hagerstown, Maryland, a group of family and friends, representing the many lives HPM had touched and enriched, gathered for the last farewell. During the funeral, amateurs around the world observed thirty minutes of radio silence.

Nine days after Maxim's passing, his beloved Jess joined him in death. She had been suffering from diabetes, and had contracted pneumonia. But the family believed that she had willed herself to die. It was as if she who had been his companion from the horseless carriage days, all down through their years, could not bear their separation. She had ridden with him in the Mark VIII. She had gone with him to Paris to translate for his ARRL, and so it was in keeping with their life together that she should continue to be with him.

The passing of Hiram Percy Maxim was a blow to the amateurs of the world. He had been their organizer and champion. He had been president of the ARRL since it was organized. He had been sponsor of QST since its first frail edition. Their April 1936 issue carried an editorial written by Kenneth B. Warner. The emotion was deep:

"Maxim was more than the presiding genius of amateur radio. He was one of the greatest men of our times... We are not alone in mourning him; many an art, many a group of doers and thinkers, both in this country and abroad, feel his loss even as we... The character of Mr. Maxim can be summed up in a few crisp words: he stood for the very highest principles in everything. He was universally respected, and no one would think of letting down so grand a chief... The organization must not be selfish; it must have orderly government in terms of majority opinion; it must work for the greatest good to the greatest number; it must not lend itself to personal ax-grinding..."

The May issue of QST carried a reprint of a tribute to Maxim published by the Hartford Engineers' Club. It stated in part:

"His death... leaves a sense of loss which a-
mounts almost to desolation. Almost, but not quite,
for with it comes the knowledge that men like HPM
never die... Few men have met life's challenge
with as high a courage as his, and there is nothing
which would mean more to him than the thought
that his passing would leave thousands of torches
which had been lighted at his... Maxim certainly
possessed that spark which we call genius, but he
had much more. HE UNDERSTOOD HIS FELLOW
MEN AND LOVED THEM."

In Clinton B. DeSoto's book on amateur radio, Two Hundred Meters and Down, there is a dedicatory preface written
by Herbert Hoover. Penned in September of 1936, it recalls
Maxim's championship of a cause to which he gave the best efforts of a lifetime:

"I well remember the battle in which Hiram Percy Maxim joined with me as Secretary of Commerce in setting apart definitely and for all time
certain segments in the radio range and dedicating
them for the perpetual use of amateurs... Mr. Maxim's sturdy mobilization of the thousands of amateurs contributed greatly to saving this field, which
has now extended into worldwide use."

Hoover goes on to praise the accomplishments of the
amateurs in both experimenting and research, as well as in
the sending of vital messages:

"Their art has added to the joy of life to literally hundreds of thousands of men, women, boys
and girls over the whole nation. Their international communications have a value in bringing a better spirit into the world.

"I consider it an honor to join in any tribute to
the memory of Hiram Percy Maxim."

DeSoto's own dedication was to the "Scientist, engineer,
author, cultured gentleman, revered by hundreds, well-beloved
of thousands, known to millions."

The Amateur Cinema League published a Maxim tribute titled "Farewell to a Chief":

"... this universally great citizen of the world
... Above all else, HPM was world-minded... his
stubborn faith in the dignity and righteousness of
the human race never faltered... he pinned his
faith to mankind above all else... he was bent on
making the world a better place to live in."

It was not only the great and the famous who mourned their Chief; every amateur in the country was stunned and saddened. The average operator in the average ham shack felt his loss deeply. One of these, Michael J. Chaveney, VE3GG, expressed the feelings of all in a poem published in the QST of April, 1936:

Across the jeweled curving dome of night
He flashed these words to me, 'Maxim-- is dead.'
     And then his key was silent.
          So was mine.

There was nothing more to say,
     Nothing we could do but listen....
Listen to that sombre lightning play
     Around the spinning globe, as ham told ham
          Our President is dead.

Slowly I drew the veil, muting the set
     'Til all the signals died, and silent
Burned the pilot light, beacon of grief,
     A candle for the dead.

Great men have died before,
     Kings and Princes.
The news ne'er moved me deep, and yet
This abyss where my heart has gone
          Plumbs all.

Maxim! Yours was the vital spark
     Which kindled for us all

Ten thousand friendships
Endeared with loves alike, exchanging keys
   To one another's hearts, and homes.

The loom you made has spun a mighty weave
Netting the whole wide world with threads invisible,
   Patterned the miracle each age so long has prayed
          for.
   Nation and creed forgotten.... as man called man
   His brother.

Henceforth this date all amateurs have marked
As yours, in silent tribute 'tween the frozen poles
   This night will muted be,
   So that the stars will wonder.

In keeping with the last stanza of this poem, on February 17, 1937, the first anniversary of Maxim's death, a Dedication Relay was held at the headquarters of ARRL. Maxim's son and daughter were present for the occasion, which was a simple exercise, with each representing station receiving a memento of the event. This was the first of these annual ceremonies, and the response was in the thousands, including one from the Governor of Michigan, proclaiming, "It was a great relay."

It was a fitting form of memorial, as the relaying of messages is one of the greatest manifestations of amateur service.

At this time, announcement was made of an annual Hiram Percy Maxim Award, by his son and daughter, "in token of Mr. Maxim's great and warm-hearted interest in the struggling experimenter." This award was to consist of a cash amount, and also a miniature reproduction of the original Wouff-Hong, which, as fable had it, was unearthed by the genial rascal, T.O.M. as the first and only known specimen of the awesome object. It has since become the symbol of the Royal Order of the Wouff-Hong, amateur secret society of ARRL conventions.

Since Maxim's great interest had been in the young amateur who was struggling under handicaps, the memorial award was not to be in the form of a contest or a competition, but a seeking out and recognition of the best young amateur under the age of twenty-one. Thus the award could be given for a specific feat in the field of amateur radio, or for the best all-around amateur record for the year, and was to include qualifications of cooperation, versatility, and well-roundedness.

Maxim's greatest devotion had been to the amateur... "amatore," lover... for the love of achievement. He loved the fraternity, the rag-chewing, the ham-fests. These were, as DeSoto put it, "an adventurous band of free spirits... who have wiped out for all time the barriers of race, language and distance... a band of good fellows, happy, convivial... playing the game for the love of it; but underneath all that, carrying on with the deep earnestness of those who have successfully pierced the veil of the unknown and garnered the secrets of science."

To fully appreciate the personality of Hiram Percy Maxim, one must comprehend the scope of the amateur. Hams are in age from eight to eighty. In educational attainments they range from grade school status to that of doctorates. Their occupations run from that of coal miners to major executives of large corporations. They include impecunious school boys, and men of great wealth. Their number includes a dedicated variety from nuns to politicians. The ham's predominant characteristic is his altruism, where he rates among the highest in the world. The amateur wants all the other hams to share in his discoveries and accomplishments, and is not competitive. He guards only his spot on the air. As soon as an amateur does something of note, he wants to show his fellow hams not only how he did it, but further, how they can do it too.

As all organizations have their creeds, formalized or otherwise, the amateurs have their code. It was written by Paul M. Segal, and is contained in the Radio Amateurs' Handbook, the Standard Manual of Amateur Radio Communication. This code has six points, and incorporates ideals of being loyal, polite, progressive, friendly, well-balanced and patriotic. That fifth point, of balance, is as unique as it is important. Most organizations promote themselves first and foremost, and all the time. Not so with amateur radio. The ham code makes it clear that radio is a hobby, and as such it must never be allowed to interfere with duties owed to home, job, school or community. In this we see the guiding hand of HPM, who believed that the amateur must never become a slave to his rig, but always its master.

After Maxim's death, his son Hiram Hamilton Maxim took over the Silencer Company. He developed silencers for submarine engine exhausts, which in World War II led to specializations for destroyer escorts, landing craft, and many navy accessory vessels. After the war the company devised a silencer for army tanks, not of cast iron, as formerly, but of stainless steel. Later, other ideas were expanded, to diversify the business. These included a successful process for obtaining fresh water from sea water.

Maxim's two autobiographical books, **A Genius in the Family,** and **Horseless Carriage Days,** had been published posthumously. The former was dedicated to HPM's son Hamilton, and the latter to Hayden Eames.

On August 5, 1938, Maxim's memory was honored by the dedication of Station W1AW, at ARRL headquarters, West Hartford, Connecticut. As the memorial tablet was unveiled, Dr. Eugene C. Woodruff, the master of ceremonies said, "It is now my great honor and privilege to dedicate this station, W1AW, to the memory of Hiram Percy Maxim... henceforth this memorial will be hallowed ground to the radio amateur." Other speakers included Hiram Hamilton Maxim, Percy Maxim Lee, and a representative of Governor Cross.

Today W1AW, Old Betsey, is enshrined at ARRL Headquarters in Newington, Connecticut. The plaque attached to this first rig is engraved to Maxim, "1869-1936... Father of organized amateur radio, beloved first president of ARRL."

The purpose of this station, which occupies seven beautiful landscaped acres, is to keep in touch with the amateurs, largely by means of bulletins distributed world-wide. There is a glorified ham shack loaded with the finest of equipment, available for the visiting ham who has his ticket with him. There

are films available to clubs, and a testing laboratory. What is developed here is described in QST for the hams to build. It is not commercial; it is not patentable. There is a technical department, a library, and the facilities for putting out the ARRL magazine, QST. Of most interest to the average visitor was the museum. Roland Bourne, Maxim's employee of many years, and fellow amateur, was the curator. He personally restored many of the ancient rigs on display in their glass cases. His workmanship earned him a first prize in restorations.

Some of the articles on display, besides the Rettysnitch and the Uggerumph, included: an electrolytic receiver of 1902, converted with a cold cream jar detector; a 1901 6-12 volt DC spark coil; an omnigraph from the early 1900's with a phone jack; a 1937 Ohio Flood rig used in a small open boat to rescue 1500 people; the pen used by Lyndon B. Johnson to sign the Reciprocal Operating Bill sponsored by ham Barry Goldwater; the first American station to work across the Atlantic; a 1932 Spanish hat, a gift of that country's hams, signed with their calls; a ceremonial book from China, inscribed, "Hand in hand, working for the same aim, there will come a day of world peace and brotherhood, your Society leading"; the original log sheet of Paul F. Godley at Androssan, Scotland, December 8, 1921; station 1KB, closed down from April 10, 1917, gathering dust till 1919, when it was put back in service, the box having been filled with linseed oil, with a valve at the bottom to facilitate draining, this rig being the dream of the early ham; a ham station, 1907 style, with homemade coils and fixed capacitors, DX 12 miles; the 1911 transmitter used by HPM, range 1 city block; and a peculiar-looking trophy, the Elser-Mathes cup to be awarded for the first American radio two-way communication from Earth to Mars.

The Hartford State Library in Connecticut has a Hiram Percy Maxim Memorial section. Articles to be seen there include: seven leather-bound notebooks of HPM's early car records and designs; many historical photographs; a framed document transmitting the Gold Medal Award of the Paris Universal Exposition of 1900 to Hiram Percy Maxim for his work on vehicles; the manuscript of HPM's three books, and many other writings; two scrapbooks of clippings giving world reaction to the invention of the silencer; and a folder of personal papers, including letters from famous personages all over the world.

On July 5, 1946 a newspaper account in the London Daily Telegraph commented on the movie version of **A Genius in the Family.** Re-named "So Goes My Love," some of the scenes were fiction rather than biography, such as the far-fetched one wherein Sir Hiram gave some phosphorescent hair oil to a friend who did his courting in the dark. The London critics rated this motion picture as "one of the most attractive comedies of its kind since "Ah, Wilderness."

On December 8, 1955, Emhart bought the Maxim Silencer Company, ending a family enterprise of almost half a century.

*   *   *

Today, across the living room of a Connecticut home, an oil portrait of HPM smiles at the painting of a pert little girl in a Kate Greenaway white dress, with big blue ribbons.

Today, every time an amateur clicks a key or sends his voice into the ether, a living tribute is being paid to Hiram Percy Maxim, the genial genius who made it all possible.

APPENDIX

Highlights of Hiram Percy Maxim's speech of January 31, 1930, before the Interstate Commerce Committee in Washington, D. C.

"Mr. Chairman and Gentlemen: My name is Hiram Percy Maxim, my residence is Hartford, Connecticut, and I represent the American Radio Relay League, the national organization of the radio amateurs of the United States....

"In considering the matter of regulating communications, as proposed in Senate Bill number six, it seems that the Senate committee on interstate commerce would do well to know certain facts regarding the amateur in radio. These facts are more important than appears upon the surface. Because the radio amateur does not represent a business or an industry, little is generally known about him. However, he is of great importance to our country, and it is for this reason that I believe this committee will be glad to receive what I have to offer it.

"Amateur radio is unique in history. Nothing quite like it has ever before existed. It is as old as radio, the great Marconi himself having started as an amateur and being truly typical of one. From the earliest days there has been something about communicating across space that has fascinated those of us who are technically inclined. There is a scientific romance to it that profoundly moves certain of us regardless of the social or financial status to which we happen to have been born. Rich and poor, educated and uneducated, old and young, with the product of our hands and our own brains, we are able to reach out into the empty ether and make contact with another intelligence."

*   *   *

"After the war [1919] the ranks of the amateurs were augmented by the thousands who had been trained in the two services. Altogether they made of amateur radio an imposing institution. This was proven in the case of various radio bills that were introduced in Congress which threatened the existence of the amateur. The tremendous improvements made in radio apparatus during the war, under the stress of war conditions, were all known to the amateurs, since they had to operate them, and thus it came about that a tremendous increase in amateur interest ensued. Their American Radio Relay League, as they affectionately term it, was put together again and the conquest of the short waves was taken up with unprecedented enthusiasm. It is difficult for me to convey to those not informed upon this subject of amateur radio the intensity of purpose of these young fellows. They unquestionably are the pick of the land when it comes to mentality and resolute character, or they would not have taught themselves the science of radio, and the telegraphic code in the first place. Furthermore their path was no easy one, for they were in the overwhelming majority of cases the sons of parents in very modest circumstances. But lack of money only whets the intensity of the amateur. One case that came to my notice is worth the attention of this committee. A certain young man, aged seventeen, in a mid-western city, was known to possess a particularly efficient station. Attention became directed at him because of his long distance records and his superior operating. Investigation disclosed the surprising fact that he was the son of a laboring man in very reduced circumstances. The son had attended the ordinary schools until he was able to work, and then he assisted in the support of the family. They were very poor indeed. Surprise was manifested that under these oppressive conditions this young fellow should have such a fine radio station. It was found that this station was installed in a miserable small closet in his mother's kitchen, and that every bit of it had been constructed by himself. This meant that such things as headphones and vacuum tubes were home-made. Asked how he managed to make these products of specialists, he showed the most ingenius construction of headphones built of bits of wood and wire. In the case of his vacuum tubes he had found where a wholesale drug company dumped its burned-out bulbs, and had picked through enough glass to blow his tubes, and enough bits of tungsten wire to make his own filaments and had literally home-made vacuum tubes... and good ones at that.

"To exhaust his vacuum tubes he built his own mercury vacuum pump from scrap glass. His greatest difficulty was securing the mercury for this pump. The greatest financial investment this lad had made in building his radio station was twenty-five cents for a pair of combination cutting pliers.

"This case illustrates the amateur spirit, a knowledge of which I consider it my duty to convey to this Committee. No explanations are necessary or called for. The case points its own moral."

*　　　　*　　　　*

"It began to be suspected that waves as short as 100 meters might be controlled. I remember vividly one of the 100 meter tests in the fall of 1923. A French amateur named Deloy had agreed to call an American amateur in Hartford, Connecticut on precisely 100 meters at precisely nine o'clock, Eastern Standard Time, on a certain Sunday evening. The receiver in Hartford was set at precisely 100 meters and the American amateur had sufficient confidence in the precision of amateur methods to make no preparations for searching around on other waves. Exactly as the second hand of a carefully-set watch indicated nine o'clock, the little signals from faraway Nice, France, began calling Hartford, Connecticut, U.S.A.

"It is difficult to explain the thrill that accompanies an experience such as this. It is sublime, and carries with it a sort of uplift that makes us better and deeper-thinking men. The precision of it all, the picture of the Frenchman sitting in his little den in France, waiting for the precise second to come around, hand on key. The Americans sitting in their little shack in a little street in New England, silently listening and watching the time... the miles and miles of lonely black ocean over which the little electro-magnetic oscillations must travel, are utterly compelling to us amateurs.

"It did not take us long to find that 100 meters was marvelously better than 200 meters. And it did not take long to find that 80 meters could be controlled and was even better than 100 meters for certain conditions. Then a way was found to keep 40 meter oscillations steady, and unbelievable records of long-distance communication in daylight were hung up. Then 20 meters was tamed and the amateurs in the Antipodes... in Australia, New Zealand, South Africa, the Philippines and Japan, were brought within reach. By 1926 the amateur on his short waves and his home-made apparatus bridged the ultimate of terrestrial distance.

"By dint of the most careful management and team-play the amateur as a whole is still constantly training himself, and those new amateurs who constantly join, to become expert radio operators, to take positions in the industry and to be available to our Government in time of need. The War Department and the Navy Department both have offered every encouragement and are fully aware of the incalculable value of the amateur. For many years not a single major break-down in general communications has occurred that amateurs have not played a major part in providing radio communications for summoning and directing relief.

"Does it not mean something that this invaluable service has been developed by the amateur voluntarily? There is nothing in any amateur's federal license that requires him to perform this service. He serves without compensation, he desires none. He works for the pure love of the thrill of doing a public service by means of his beloved radio. Is this 100 percent altruism or no? Is it worth preserving, or no?

"Permitting those of our young men who have the natural attainments, regardless of their status in life, to pursue amateur radio, has contributed to the placing of our country in the present position at the head of the radio communications art... it added to our national wealth in providing encouragement and opportunities to thousands of young Americans who probably would never have possessed them otherwise. As their leader,

I appear before you now, in 1930, as I have appeared many times before, to plead that at all hazards you look out for the amateur of today and also for those amateurs who are to come.

"We have no comment to make upon the wisdom of the provisions of SB6. We wish merely to point out that whether it is a Radio Commission, or the Department of Commerce, or a Communications Commission that is given jurisdiction over radio, provision should be made for the amateur.

"We have fared very well under the administration of the 1927 law by the Federal Radio Commission and the Radio Division of the Department of Commerce. They both appreciate the value of the radio amateur. The Federal Radio Commission has put at our disposal the maximum of the facilities made possible for radio amateurs by the 1927 international treaty, and in a recent analysis of the high-frequency spectrum publicly expressed regret that the treaty so limited it that it could not continue the previous more extensive allocation of channels to this important group, the radio amateur. We shall be very well satisfied if the government communications commissions of the future deal with us as fairly as have the existing agencies.

"However, any governmental regulative agency will be besieged by commercial interests to grant ever greater numbers of channels to them. It will be urged upon such an agency that the amateur channels are worth thousands of dollars in earning ability. Let me entreat that you believe that those few radio channels which have been left the amateur constitute a value to our nation incalculably greater than any possible money earnings that could conceivably be developed from them by any commercial company.

"Finally, in considering a means of regulating communications let me urge you with all the force and sincerity I can command that you bear in mind the radio amateur and make possible his continuance. To do less would be nothing short of a national catastrophe."

Author's Note: As a result of this speech before the U.S. Senate Committee, the bill was killed, and amateur radio was saved, for the amateurs of that day, and for posterity.

* * *

More great publications from the Ham Radio Publishing Group...

NOVICE RADIO GUIDE
by Jim Ashe, W1EZT
Complete handbook for the beginning Amateur. Theory, putting
your station together and operating. 144 pages.
Order HR-NR $3.95

HAM NOTEBOOK VOLUMES I and II
by Jim Fisk, W1HR
Collection of the very best from HAM RADIO Magazine's Ham
Notebook column. No repetition between volumes. Volume II con-
tains projects and ideas never before in print.
Order HR-HN1   Volume 1 $3.95
Order HR-HN2   Volume 2 $4.95

LOW & MEDIUM FREQUENCY SCRAPBOOK
by Ken Cornell, W2IMB
Informal scrapbook exploring the 160 - 190 kHz frequencies — get
on the air legally without a license using simple, easily built
equipment. Full info regarding FCC rules and regulations.
Order HR-LF $6.95

SECOND OP
by W2AB (formerly W9IOP)
Super handy DX operating aid telling the prefix, continent, DX
zone, country name, local beam heading, time differential, QSL
bureau and postal information.
Order HR-20P $3.50

SATELLABE
by Kaz Deskur, K2ZRO
Sophisticated, two-color circular slide rule for tracking sun-
synchronous Satellites. Calculate azimuth and elevation information from the user's location for that entire orbit and all subsequent
orbits.
Order HR-STL $7.95

AMATEUR SINGLE SIDEBAND
by Collins Radio Company
The bible on Amateur single sideband. Introduction to SSB, nature
of SSB signals, exciters, R-F Linear Amplifiers, SSB receivers, test
and measurements and what comprises the Amateur SSB station.
Order HR-SSB $4.95